M000213751

HERMANN MEMORIAL LIBRARY
SULLIVAN COUNTY COMMUNITY COLLEGE
112 COLLEGE ROAD
LOCH SHELDRAKE, NY 12759

Seeds of Resistance,
Seeds of Hope

Seeds of Resistance, Seeds of Hope

Place and Agency in the Conservation of Biodiversity

Edited by

VIRGINIA D. NAZAREA,
ROBERT E. RHOADES, AND
JENNA E. ANDREWS-SWANN

THE UNIVERSITY OF
ARIZONA PRESS

TUCSON

333
95316
SE32N

The University of Arizona Press
© 2013 The Arizona Board of Regents
All rights reserved.

www.uapress.arizona.edu

Library of Congress Cataloging-in-Publication Data
Seeds of resistance, seeds of hope : place and agency in the conservation of
biodiversity / edited by Virginia D. Nazarea, Robert E. Rhoades, and Jenna
Andrews-Swann.
    pages cm
  Includes bibliographical references and index.
  ISBN 978-0-8165-3014-4 (hardback : alk. paper)
    1. Agrobiodiversity conservation—Cross-cultural studies. 2. Food
security—Cross-cultural studies. 3. Local foods—Cross-cultural studies.
4. Seeds—Cross-cultural studies. I. Nazarea, Virginia D. (Virginia
Dimasuay), 1954–, editor of compilation. II. Rhoades, Robert E., editor of
compilation. III. Andrews-Swann, Jenna, 1980–, editor of compilation.
  S494.5.A43S39 2013
  333.95'316—dc23
2013011198

Publication of this book is made possible in part by the proceeds of a permanent
endowment created with the assistance of a Challenge Grant from the National
Endowment for the Humanities, a federal agency.

Manufactured in the United States of America on acid-free, archival-quality paper
containing a minimum of 30% post-consumer waste and processed chlorine free.

18 17 16 15 14 13   6 5 4 3 2 1

This volume—simmered in the love of collard greens, black-eyed peas, ham hock, corn bread, the American South, and all the world's resilient Souths—is dedicated to Bob Rhoades.

CONTENTS

II. Agency and Reterritorialization in the
Context of Globalization

PREFACE

Four words that are not normally used, or not used often enough, in relation to the conservation of plant genetic resources in particular and biodiversity in general are "resistance," "hope," "place," and "agency." Why focus on these instead of, say, sustainability, rationality, economic value, and political strategy?

Conservation is all too commonly viewed as something that requires intentionality—a ponderous matter people must deliberate on and methodically execute. Worse yet, it is often regarded as what more developed, formally trained, or well-positioned people ought to get less developed, informally educated, and poorly positioned people to think about and do. Environmental management—or its earlier and more lyrical rendition, stewardship—has so colonized our mindset that we often forget that it is a fairly recent invention. Before the conceptual separation of mind and body, and of humans and nature, and before we were preoccupied with dissecting the objective from the subjective, people were part of a highly interactive and largely animate environment on which they depended not only for survival but also, more important, for completion. When the surrounding mountain range is your "mother," the gnarled tree is your "grandfather," striped stones are your "scouts," newly harvested potatoes are your "infants," and tangled beans are your "sisters," you do not distance yourself to manage and protect. Indeed, for many cultures that continue to dwell in an intimate-animate landscape, such separation and hierarchy are inconceivable, if not downright distressing.

This message was brought home to us when, preparing a compilation of myths and legends in Cotacachi, Ecuador with local collaborators, it became increasingly clear that there was no Quechua equivalent of "conservation," "management," or even "stewardship." As indigenous leaders Magdalena Fueres and Rafael Guitarra politely yet emphatically pointed out, the principles *igual y igual* (among equals) and *no paternalismo* (no paternalism) applied not only to our relationship to them as outsiders who bring in funding for research projects but also, in their world, to the relationship between humans and other-than-human nature. Does conservation have to be intentional to qualify as such, or is it the sum and synergy of what people do in their daily lives out of caring and remembrance?

For all its hegemonic power, globalization has occasioned a willful—at times whimsical—reterritorialization, a reinforcement of threatened senses of identity and place. Granting that the modern world is not given to greeting "sister bean" or "baby potato" warmly on the supermarket shelf, there remain places where biodiversity is in fact nurtured and defended. We see this in the expanding web of emotional attachment to the land, to rural places, mindful lifestyles, and wholesome food. In affluent urban areas, specialty or niche markets, local farmers' markets, nouvelle regional cuisine, and rustic gourmet restaurants appeal to what one airline magazine writer so aptly dubbed our "memory palate." The more cosmopolitan the population, the more intense social movements such as slow food, local food, and heirloom crops and small breeds conservancy are in defending the "ark of taste," the "locally grown," the "grass fed," and the "old-timey." It is likely that they fill a real need, enabling people to preserve and savor different tastes that signify connections and memories where these might be fragile and tenuous, or virtually nonexistent. However, outside of the public sphere, we also witness this kind of attachment in more intimate marginal spaces like homegardens and kitchens where people quietly tend to what they enjoy, and what gives meaning to their lives and their relationships. Somewhere(s), modernity and globalization have run aground. Whether rooted or transported, people, plants, and places may have been bent and bruised but, on the whole, have remained remarkably pliant and disarmingly plural.

Without denying the gravity of erosion and loss of biological and cultural legacies, we need to acknowledge the countervailing forces of resistance and hope, first, because, as contributions to this volume show, they do exist; second, because they are often overlooked; and third, because they offer inspiration and vital clues to resilience and sustainability. Place-based biodiversity conservation emanates from sensory and affective engagement with the myriad living complexes in our worlds. Like breaking the soil, sowing seeds, and harvesting grains, this engagement is individual and personal on one level but also collective and communal on another, and for this reason broadly compelling. Still, place making and resistance to homogeneity are possible only when people believe they have a modicum of agency—that to some extent they can influence the course of things and events—and some basis for hope. As academics, practitioners, and activists, we cannot afford to ignore this aspect of conservation, especially as we attempt to reverse our priorities from *ex situ* collection and cold storage of plant genetic resources to the return and reintegration of these priceless legacies to their original custodians and their conservation *in situ* as well as *in vivo*.

If science and advocacy are to find ways to strengthen the earth's life support system, we must first understand the strong yet subtle *ties that bind*. Only then can we explain, predict, and influence the course of biodiversity on this planet. On purpose, multiple voices are included in this volume. They cross or straddle disciplinary, generational, national, and political borders. The contributors demonstrate the importance of cultural memory in the persistence of traditional or heirloom crops and varieties, as well as the agency exhibited by displaced and persecuted peoples in place making and reconstructing nostalgic landscapes, primarily in the form of gardens from the homeland or from the past. They explore the mechanisms by which intrusions from agricultural extension, markets, and genetically modified organisms (GMOs) are countered, as well as the dynamics of local initiatives in repatriation and *in situ* conservation. They examine the conservation of biodiversity at different scales, from different perspectives, and with different theoretical and methodological approaches and debate the direction biodiversity conservation needs to take in the future. This eclectic collection brings to the table a variety of lenses and tools from anthropology, sociology, genetics, plant breeding, education, advocacy, and activism that can be brought to bear on a number of pressing problems we face in shifting our conservation paradigm from centralization and control to complementation to facilitation.

We acknowledge all the participants in the international and interdisciplinary workshop "Seeds of Resistance/Seeds of Hope: Repatriation and *In Situ* Conservation of Traditional Crops," held in Athens, Georgia, on April 30–May 1, 2004. In the in-between years, as questions were raised and debated long-distance, interspersed with site visits and face-to-face interactions in incipient and long-term collaborative projects, the chapters for this volume evolved and meshed. This volume, then, is but one product of that workshop, one that is meant to stimulate further scholarship and action. The chapters included here have been mulled over and rewritten to reflect more current developments and concerns. As well, entirely new contributions reflecting the kind of collaborations we espouse have been recruited. The whole enterprise would not have been possible without the support of the Sustainable Agriculture and Natural Resources Management Collaborative Research Support Program (SANREM CRSP), the State Botanical Garden of Georgia, and, at the University of Georgia, the Institute for Behavioral Research, the Wilson Center for the Humanities and Arts, the Office of Public Service, the Department of Anthropology, and the Ethnoecology/Biodiversity Laboratory.

Our appreciation goes to the Office of the Vice Chancellor for Academic Affairs at the University of Georgia for coming up with a faculty development program as innovative as the Study in a Second Discipline, from which we greatly benefited (Nazarea in Law and Rhoades in Geography and Climate Change). Our one year of coursework and intellectual exchange with other faculties on campus sensitized us to the distinct and at times unnerving logic and idioms of different fields, which can be a deterrent but also a spark. The heightened awareness that resulted from this exchange opened our minds to possibilities for cross-disciplinary inquiry and planted the seed for the conference and this volume.

We would like to take this opportunity to express our gratitude to our colleagues at the International Potato Center (CIP) in Lima and the Association for Nature and Sustainable Development (ANDES), as well as the Quechua communities comprising the Potato Park in Cusco. The intellectuals behind the cultural revitalization movement in Peru, notably the Andean Project for Peasant Technologies (PRATEC), the radical agronomists, and the progressive intellectual property lawyers likewise provided many hours of stimulating conversations. In Lima and Cusco, we have spent some of the finest and most memorable times of our professional and personal lives, and we are indebted to our hosts for the collegiality and hospitality we received. Finally, we thank our former and present students who facilitated the initial and ongoing exchange, namely, Maricel Piniero, Milan Shrestha, Shiloh Moates, Juana Camacho, Adam Henne, and coeditor Jenna E. Andrews-Swann. After the conference, Juana Camacho, Kristine Skarbø, and Madalena Monteban joined us in further explorations of various topics and issues at the Potato Park and at CIP. The seeds of resistance and hope have undoubtedly found fertile ground.

This volume celebrates the life and works of Dr. Robert E. Rhoades. Although it is unusual to dedicate an edited volume to its coeditor, this is our way of acknowledging his inspiration, collaboration, and support. He pioneered the farmer-back-to-farmer approach that emphasizes the need to start, *always*, from the farmer's point of view and to cross-refer any scientific innovation to local knowledge and constraints. He dedicated his life to breaking down the distinction between academic and applied research as well the demarcation between academic and popular publications in anthropology and agriculture. His teaching and research successfully integrated theory and action; for him, all anthropology is applied, and all anthropology is public. Two of his most recent publications—an edited volume on landscape-lifescape approaches to natural resource management in Ecuador, *Development with Identity: Community, Culture, and Sustainability in the Andes* (2005), and a semiautobiographical book on the trajectory of mountain research that incorporates a new agenda for sustainable mountain development, *Listening to the Mountains* (2007)— demonstrate the depth of his understanding and commitment.

Among the many hats he wore, he liked the "Professor Rhoades" and the "Farmer Bob" hats best. In demonstrating the arbitrariness of these boundaries, he challenged not only taken-for-granted categories but also, more fundamental, hierarchies that are all the more formidable because they have the appearance of being set in stone. His many path-breaking contributions to both fields, his publications in popular press that have been reprinted numerous times in scholarly volumes, and his agrarian populism that continues to motivate cohorts of his students who are now mentoring and inspiring their own are testaments that it can be done. Whether in the realm of agricultural development, indigenous knowledge, agrobiodiversity, climate change, or landscape restoration, Bob Rhoades's legacy in the defense of people, plants, and places lives on.

Virginia D. Nazarea

Seeds of Resistance,
Seeds of Hope

Conservation beyond Design

An Introduction

VIRGINIA D. NAZAREA AND
ROBERT E. RHOADES

We cannot begin to talk about the conservation of biological diversity without first taking account of legacies, traces, and tidemarks, for conservation rarely begins with an external program or a streamlined design. On the contrary, it begins with genetic and cultural heritages in different degrees of vitality and disrepair, interconnected reservoirs of alternatives that draw on and replenish each other, and ideologies and encounters that either enrich or decimate. So, we begin by asking, does biodiversity conservation have to be intentional and preconfigured in order to count as such? Or is it the haphazard product of what people do in their daily lives based on more or less unremarkable remembrance, resistance, and resilience? Do these *unprogrammed* and *undesigned* instances of on-the-ground conservation go by the wayside as idiosyncratic blips, or do they add up, challenge hegemonies big and small, and amount to a compelling countermovement?

A significant redirection has been going on in the conservation of biodiversity. Earlier approaches sought to collect and centralize plant genetic resources *ex situ*, away from where they originated, in such repositories as botanical gardens and genebanks. In the past two decades, questions about deterioration of genetic material, on one hand, and freezing of their evolutionary potential, on the other, necessitated complementing

3

standard *ex situ* strategies with more experimental *in situ* approaches such as protecting diversity-rich "hot spots" and connecting them through biological corridors or conducting systematic "on-farm" crop conservation trials, identifying synergies of environmental and cultural factors that foster biodiversity and subsequently replicating positive outcomes (Brush 1991, 2000; Pimbert 1994; Prance 1997). In the case of agrobiodiversity—or the genetic diversity of the world's important crops—the urgency of *in situ* conservation was further driven by more ethical and political dilemmas of sovereignty over germplasm, along with issues of access to and benefits from their use (see chapter 10, this volume).

The overall paradigm shift in plant genetic resources conservation started in the mid-1990s in response to scientific critiques of *ex situ* conservation (Maxted et al. 1997; United Nations Environment Program/ Convention on Biological Diversity 2001). The argument centered on two main observations pertaining to accessions in genebanks: (1) deterioration of genetic material due to imperfect storage conditions and human error, leading to changes in gene frequencies, and (2) freezing of evolutionary potential due to long-term storage, resulting in the inability of germplasm to respond to environmental changes. Long-term storage under such artificially controlled conditions as prevails in genebanks can render the collections unstable, archaic, and irrelevant after some time (for more exhaustive treatment, see Wilkes 1991; Cohen et al. 1991; Hodgkin et al. 1995). While on the surface contradictory, deterioration and freezing both suggest that *ex situ* conservation, like any ameliorative measure, has some serious limitations. Claims and counterclaims regarding ownership, rights, and responsibilities in relation to plant genetic resources and associated knowledge only complicated this scenario. The declaration of national sovereignty over plant genetic resources in the Convention on Biological Diversity resolved some significant questions but spawned others (Fowler and Mooney 1990; Kloppenburg 1990; Dutfield 2000, 2004; Carrizosa et al. 2004).

In tandem, academic research and international programs began focusing attention on *in situ* conservation of landraces, including native varieties and wild relatives, in places that are either characterized by outstanding species richness or under serious threat of genetic erosion. Experience, however, has shown that conservation in nature reserves and farmers' fields is considerably more unpredictable and intractable than *ex situ* conservation under tightly controlled conditions. In other words, *in situ*—and even more so, *in vivo*—conservation rarely yielded to any programmatic design and is, on the whole, idiosyncratic or, one might say, upbeat. We have found, for instance, that to try to impose discipline and order in homegardens in an attempt to extract universal

principles and come up with a succinct list of "lessons learned" is to suf-
focate the very spirit (or, more simply put, joy) that nurtured diversity
in the first place (Rhoades and Nazarea 1998; Nazarea 1998, 2005). If
there is any lesson learned, it is that biodiversity flourishes under conditions of marginality, hand in hand with memories that enliven culinary
and healing traditions, as well as reciprocity and commensality. Hence,
allowing spaces for traditional ways of life to prosper *in vivo* as viable alternatives to global monocultures of capitalism and consumerism is even
more imperative than the collection of germplasm and codification of
traditional knowledge in genebanks and archives (see also Hunn 1999;
Ingold 2000).

Uncontrived and unexpected, this other trajectory has slipped by our
consciousness largely unnoticed, and thankfully so, or it might have
been prematurely nipped or disciplined for being too messy. It goes beyond seeking to complement *ex situ* with *in situ* approaches and touches
the nerve of conservation on the ground, or conservation with a small *c*.
Examples can be found in diverse plantings of varieties favored for home
consumption on borders of commercial plots, in tangled fields of greens
and old-timey orchards, in women's homegardens, and in immigrants'
reconstructed landscapes. Moreover, beyond the field and the garden,
conservation of biodiversity is reinforced by the longing and demand for
local food and for comfort food from one's homeland (however defined)
or from the past. In acknowledging, facilitating, and reinforcing viable
complexes that are embedded in the idea of place, accessible to local
custodians, steeped in cultural significance and emotional attachment,
and open to evolutionary change, living or *in vivo* conservation constitutes a quiet revolution—one built on agency and offering hope.

The emergent problems of biodiversity conservation, whether *ex situ*,
in situ, or *in vivo*, are multifaceted and complex. Not to be ignored, particularly where staple crops are concerned, is the dearth of agrobiodiversity that can still be conserved where the diverse traits originated, the
very definition of *in situ* conservation (Plucknett et al. 1987; Norgaard
1988). Many important crop species and varieties have been abandoned
due to pressures from agricultural development and market integration.
Logically enough, these international research and development efforts
were concentrated in centers of diversity and domestication (Zimmerer
1996; Brush 2004). So the question as we see it is, where or, more precisely, how does one begin to replenish what conceivably has been "lost"?
And how can local loss make any sense in the context of global conservation in genebanks, botanical gardens, research stations, and biological
corridors? To complicate matters further, how do we take into account
the movement of seeds in connection with global flows of people and
commodities—the serial displacement and relocation of people and

their plants that create multiple repositories of biodiversity and memory *elsewhere*? Conceptually, how is the landscape of loss to be reconciled with the landscape of memory?

The radical change in paradigms of plant genetic resources conservation requires a commensurate qualitative leap regarding our theoretical and methodological approaches. We cannot proceed with the same old conceptual frameworks, rigid disciplinary boundaries, and unquestioned political positions, however useful, effective, or safe they may have been in the past. To forge new directions in plant genetic resources research and conservation, there is a critical need to understand the mechanisms and dynamics of *everyday conservation in place*. In this volume, we offer some leads in the form of novel combinations of research approaches and collaborations. The aim of these methodological forays is to glimpse the subjective even as we try to grasp the objective, for only then can we begin to understand what actually moves and sustains people as they relate to and deal with the diversity of their human and other-than-human kin. In the domain of advocacy and action, the studies presented here examine cases of marginality and resistance in the form of heirloom gardens and orchards, on-farm conservation and community genebanks, memory banks and communal fields of ancestral futures, and immigrant or *trans situ* conservation. This kind of recrafting in methodology and application enables a fundamental reorientation that is particularly important in facing the increasing demand for the repatriation of landraces deposited in *ex situ* collections—well over half a century's worth (Vavilov 1987)—back to their former places, a prospect likewise besieged by seemingly endless claims and contradictions.

Repatriation of traditional varieties may be initiated by genebanks or demanded by local people, usually a combination of both. Calamities like earthquakes and floods, for instance, prompt a response from international agricultural relief agencies to bring in new supplies of seeds and thereby restore livelihoods to affected communities (Buruchara et al. 2002). In other cases, as in Cusco, Peru, local farmers organize with assistance from nongovernmental organizations (NGOs) and approach genebanks to initiate the repatriation of their germplasm (see chapters 1 and 12, this volume). These germplasm materials have been characterized in international development spheres as *seeds of hope*, but they are at the same time *seeds of resistance* because of the powerful will and sustained action on the part of disenfranchised populations to retrieve "lost seeds" and reintegrate them into their agriculture (see chapter 1). By the same token, repatriation validates local sovereignty and triggers the revival of culinary traditions, patterns of social exchange, and property regimes that have long been forgotten or lain dormant (figure I.1). Repatriation is an act that is engaged and affective; it constitutes a social

FIGURE I.I. Circle of Elders for Repatriation Priorities held in Sacaca, Cusco, Peru. Photo by Robert E. Rhoades.

movement radiating from the heart out and from the ground up (Nazarea 2006).

We can better understand the process of repatriation and *in situ* conservation when we realize that traditional crops not only have been "lost" because of climatic, market, and political upheavals but also in many cases have been systematically siphoned off in the name of conservation, world hunger, and, although somewhat more covertly, commerce and industry (Gonzales 2000; Ehrlich 2002; Altieri 2003; Mgbeoji 2006). In addition, many native varieties have been displaced by the introduction of modern cultivars of the Green Revolution, the so-called miracle or high-yielding varieties of crops, and a tighter integration into markets (Shiva 1993; Nazarea 1995, 1998; Fowler and Hodgkin 2004). An even more recent and insidious cause of "loss" comes from contamination of native crops right where they are, oftentimes with genes from genetically modified organisms, or GMOs, that have been introduced to improve agriculture and food security (Celis et al. 2004; Engels et al. 2006; Scurrah et al. 2008; Chandler and Dunwell 2008). Thus, to repatriate is to return what has been lost, displaced, or contaminated, from collections in genebanks that hold these irreplaceable materials in trust and for posterity, back to their original habitats and custodians. Repatriation coupled with *in situ* and *in vivo* conservation reverses the post-Enlightenment flow of germplasm and replaces a paradigm based on expropriation and control with one that is based on cultural integrity and revitalization (figure I.2).

This collection of chapters engages a number of theoretical and practical intersections of culture, policy, and science in the conservation of biodiversity. Foremost is the role of human agency in the nurturance of plant genetic diversity. While cognizant of political, economic, and social constraints with deep historical roots and long, powerful branches, the contributors nonetheless recognize some agency, or *room to maneuver*, that allows for outright resistance to homogenizing forces in some cases, independence and nonchalance in others, and a sensuous reclamation of place in still others. All these modes of countering homogeneity-in-modernity involve human sovereignty and resilience in marginal spaces where a sense of place can be elaborated within a milieu of memory. Our concern here is how this place-based conservation can be honored and buttressed against the displacement and alienation concomitant with globalization. What comes out again and again, however, is that much of the power of place-based biodiversity conservation is internally derived and internally driven. Enabling combinations of intervention and support simply allows what is already there to flourish.

FIGURE I.2. Local ANDES (Associacion para la Naturaleza y el Desarollo Sostenible) [Association for Natural Resources and Sustainable Development] *técnicos* preparing the land using a traditional foot plough, the *chakitaqlla*. Photo by Robert E. Rhoades.

Agricultural research and development in land grant institutions, national programs, and international centers are guided by parsimony and efficiency. Programmatic design is often the standard, and universality the rule. Likewise, in the academe, it is difficult to avoid the disciplinary microcosm that breeds the "idols of the tribe" that we rarely question. Nor do we often confront the possibility that the models we congeal in our various disciplines are heuristic distillations, perhaps even elegant distortions, and not necessarily accurate representations of the phenomena we are interested in. Socialized as we are to be good "tribespeople," it is difficult to set aside our disciplinary lenses to appreciate other frameworks and contributions. That being said, might it be possible to cultivate innovative configurations of research, development, and advocacy that permit creative openings and meldings? We believe this can be accomplished when sectoral and disciplinary borders are more porous and "crossable." We hope this volume will demonstrate that, unsettling as it can be, cross-fertilization can challenge complacent thinking and enable us to perceive with greater clarity and depth.

The contributions to this volume document and analyze various cases of resistance, resilience, memory, and dwelling—whether in the American Southerners' creative saving and passing along of heirloom apple trees, in the Japanese refusal of foreign, genetically engineered soybeans, or the Mayan effort to block contamination of their Mother corn—and attest to the irrepressibility of human will and whimsy. In the spirit of crossing borders, various lenses are called to bear on confounding problems and prospects we face in biodiversity conservation in relation to territorial rights and intellectual property rights as these play out, for example, in the *zoteas* (elevated gardens) of the Pacific Chocó of Colombia, in the communal gardens and genebanks in Ecuador and Peru, and as emergent forms of *trans situ* conservation by refugee and immigrant gardeners reconstructing nostalgic landscapes of the global South in much of the global North. The scientific documentation and analysis of these cases of self-determination are important, but we need to pay equal attention to the contributions of advocacy and activism where scientific shoes fear to tread. Here we try to elaborate a space where this kind of conversation can take place, a space where we are emboldened to look our idols in the eye and lay down our intellectual shields in order to see, feel, smell, taste, touch, and defend the proliferation of difference that leads to conservation of biodiversity.

We have organized the volume into part 1, "Marginality and Memory in Place-Based Conservation," and part 2, "Agency and Reterritorialization in the Context of Globalization," to focus attention on some of the more elusive dimensions of seeds of resistance and seeds of hope. In highlighting memory and embodiment that bring about the continued cultivation and defense of biodiversity, we draw our inspiration from men and women of varying degrees of locality and autonomy who have kept diversity in place, or transported it from place to place, and whose quotidian sentiments and actions add up, elaborating a form of conservation that is more sensuous than rational. We believe that in the present climate of denial and/or dismay about the threats to biodiversity and the prospect of irreversible erosion and loss, the set of perspectives in this volume can shed some light on the creative forces at work or, at the very least, provoke productive rethinking of our stalwart paradigms.

In the opening chapter of part 1, Virginia D. Nazarea argues that despite claims that biodiversity is decreasing worldwide at alarming rates, local populations, indigenous groups, and others who find themselves marginalized in our globalizing world are in fact resisting this purported decline. She considers the role countermemory plays in three Peruvian life histories and concludes that repatriation of traditional crops rests on people living in a milieu that allows some autonomy and remembrance. In chapter 2, Susannah Chapman and Tom Brown document the existence of an informal network of "old-timey" apple growers that has managed to recover and "re-replace" apple diversity in the American South. The contribution of these apple keepers in finding and propagating old apple cultivars and their dedication and ingenuity in keeping these heirloom fruit trees in circulation illuminate the ways in which the memory of apples is bound to social relationships, senses of place, and, ultimately, persistence of biodiversity. Along a similar vein, James Veteto and Kevin Welch found that although the spread of mainstream American foodways has resulted in significant erosion of the diversity of their traditional plant varieties, the Eastern Band of Cherokee still retain a long and varied inventory of favored food plants, both wild and cultivated. Chapter 3 describes how their culinary traditions and heirloom seeds exist in pockets throughout the Eastern Cherokee reservation, signifying their continued relevance and resilience.

Picking up the thread of significance of sense of place in indigenous resistance, Tirso Gonzales elaborates indigenous, place-based conservation and contrasts the "culture of the native seed" with the "culture of the commercial seed." In chapter 4 he argues for greater holism in *in situ* conservation and calls for the decolonization of the contemporary worldview. Magdalena Fueres, Rodrigo Flores, and Rosita Ramos present an indigenous perspective on seed saving from Cotacachi, Ecuador, in chapter 5 and describe a collaborative project, the Farm of the Ancestral Futures, which was initiated by a group of indigenous women to reverse the genetic erosion of Andean crops. The story that unfolds demonstrates how an undertaking aimed at the conservation of agrobiodiversity can simultaneously recuperate the knowledge of elders and make young people aware of the value of their legacy. Wrapping up on the problem of land and territory in the Americas, Juana Camacho examines Afro-Colombian women's contributions to the reconstruction of the social and cultural fabric and to the transmission of ethnobotanical knowledge. As she points out in chapter 6, the women have lost access to the land and resources that supported their livelihoods, social relations, and identity, yet homeplace making persists in a context of ethnic reaffirmation.

In part 2, using interviews, speeches, and his personal experiences working with the Maya Seeds of Resistance, Peter Brown in chapter 7 presents the Zapatista initiative to safeguard indigenous varieties of corn from being contaminated by transgenic introductions. He describes the ways the farmers in Chiapas are actively conserving native corn in a communal seed bank and promoting the adoption of the seeds in safe havens in other parts of the world. Also on the receiving end of the products of genetic engineering, Japanese consumers have employed various cultural strategies to resist the invasion of foreign, genetically engineered soybean varieties on several fronts. In chapter 8, Richard Moore examines contemporary efforts in Japan to preserve soybean diversity as manifested in local soy varieties and processed soy foods that were developed in different regions in response to local food tastes and environmental conditions.

Cary Fowler departs from the theme of local defense of plant genetic resources and asks at what point the Convention on Biological Diversity becomes an impediment to conservation and development. Emphasizing the worldwide interdependence on the free and unimpeded exchange of crucial crop genetic resources, in chapter 9 he explains how dominant legal conventions governing biodiversity can adversely affect access and food security. He calls for a reexamination of scientific and political ends and means to avert a global impasse that will be detrimental to all stakeholders. In chapter 10, Kristine Skarbø considers the multiple and divergent meanings for different stakeholders of three concepts integral to international treaties and regulations: biodiversity, right, and access. She investigates the situated meanings of these key concepts using photovoice and focus group discussions participated in by scientists, NGO workers, and farmers in Cusco and Lima, Peru, and points to possible directions for seeking some consensus on these pivotal concepts.

With globalization, neither concepts nor seeds are fixed in time and space, and every deterritorialization provokes a reterritorialization, as the final chapters demonstrate. Jenna E. Andrews-Swann looks at the cultural reconstruction of landscape using case studies of Cuban immigrants living in Miami, Florida, and in rural Moultrie, Georgia. In chapter 11 she argues that community events, religion, food, and gardens combine in a uniquely Cubamerican landscape that reflects a selective sense of nostalgia for a "lost" homeland. Members of the diaspora "re-member" and remake a sense of place in order to maintain a connection to the island when they cannot (or will not) travel there themselves, resulting in resistance to both political oppression and cultural homogenization. In the concluding chapter, Robert E. Rhoades examines the dynamics of recovery, repatriation, and reterritorialization among three globalizing populations: Southern U.S. heirloom seedsavers representing a marginal population within the global core, Vietnamese immigrant gardeners in the United States transporting and maintaining their native plants in an alien land, and indigenous farmers of Ecuador struggling to recuperate their ancestral crops and knowledge in a center of crop origin and diversity. He concludes that although globalization seriously threatens agrobiodiversity in all cases, localized responses based on everyday attachment to place, foodways, memory, and identity form powerful countervailing forces to homogenization and genetic erosion.

Development, politics, and cataclysmic events are implicated in the destabilization of cultural and biological legacies, particularly of subordinate groups. And yet, we have witnessed numerous challenges to wholesale erosion and loss. It is, as a matter of fact, countered at every turn not only by intervention measures but also, more effectively and more poignantly, by human agency at the margins. Seeds of Hope projects such as those in Rwanda, Cambodia, Nicaragua, and Honduras have attempted to rehabilitate the decimated seed stocks of farmers affected by natural calamities or political instabilities. The repatriation of *papas nativas* (native potatoes) or *papas rusticas* (rustic potatoes) as "seeds of resistance" in the Peruvian Andes has facilitated the recuperation of concrete and ingestible symbols of revitalization of cultures and places. In Ecuador, the reintegration of *semillas ancestrales* (ancestral seeds) in individual and communal fields in highland Cotacachi from the national genebank in Quito honors the elders and motivates the youth in their communities. In each of these cases, however, what is striking but often glossed over is that, wherever they were and in whatever circumstances they found themselves, the farmers had held on to some of their seeds and seed exchange networks in the face of agricultural extension, expanding markets, civil strife, and natural disasters. Although, for the most part, they welcomed external assistance, they had their own cultural and biological reserves to draw from. From our point of view, the greatest tragedy would be to wipe out all the tracks—all pockets of diversity and trails of memory—that enable people to get back in place.

Just as conservation does not emanate from a streamlined design, repatriation is rarely, if ever, done on landscapes devoid of any remnant of memory or choice. Even though threats and outright affronts to local people and biodiversity are everywhere apparent, repatriation is overlaid on landscapes that are intimate and animate, dwellings replete with recollections and associations. Farmers hold in place as many of their favored and familiar crops and varieties as they can keep, if not in their commercial fields then along the borders and in homegardens. In cases of displacement and immigration, the homeland is carried in concealed seed packets that, once a destination is reached, become indispensable tools for authoring and anchoring place. We believe that it is unlikely that legal, economic, and sociopolitical integration will ever completely sap these marginal spaces of their viability and strip them of their sovereignty; even refugees resist. However, we are not suggesting that this is reason for complacence. Instead, we see it as reason for starting fresh and thinking through our premises and assumptions. If we dismantle our rational frameworks and do away with our formulaic approaches, what are we left with? Can we imagine knowledge systems that are mutually respectful and permeable, social movements that are not primarily political and angry but sensual and celebratory? Can we spend more time in listening than in positioning, dwelling a little bit more and controlling a little bit less?

Perhaps the local food movement that is currently capturing imaginations and firing passions worldwide can offer us some leads. This sweeping phenomenon starts from alienation from the uniformity and blandness that are part and parcel of modernity. We see how a now-pervasive social movement is anchored in something as prosaic—meaning tangible, edible, and, one might add, perishable—as produce, grains, and meat and how rationalities and subjectivities all get mixed together and "cooked" in a highly potent broth. The effect is both counterintuitive and contagious on many fronts; how else can one explain the willingness to pay double for an heirloom tomato at a farmers' market or to spend the time, money, and energy to grow some greens in an urban space? The widely felt need for reembedding a depersonalized food system becomes both more nuanced and elusive when we venture into the more private realm. And venture we must, for in order to endure, these sentiments need to be embodied; in other words, they have to come to roost in bodies, hearths, gardens, and gathering tables. *Dwelling is in the details.* From here, it is but one step to understanding how biodiversity conservation, to be as enduring as it is compelling, has to be *placed* in what holds some meaning, and what makes some sense, in people's everyday lives.

A conversation on *ex situ*, *in situ*, and *in vivo* conservation is going to be very difficult to carry out from the purity of a single disciplinary standpoint. It entails maintaining one's disciplinary grounding but being both daring and humble enough to cross borders, including those of development and conservation practitioners and activists who are working with local people on the ground. Negotiating this new terrain requires a great deal of suppleness as well as the inclusion of many voices—from the disciplines of biology, anthropology, law, and economics; from the vantage points of science, policy, and advocacy; from the "takes" of seasoned researchers and young scholars; from the "stakes" of indigenous peoples and the NGOs they choose to work with; and from plans and agendas and the sectoral, national, and international interests they represent. The voices included here are local, global, and transnational, and they seek an honest conversation about place, about agency, and about hope. Together, they honor and call not for a conservation without design, but for a conservation beyond one.

References

Altieri, Miguel A. 2003. The sociocultural and food security impacts of genetic pollution via transgenic crops of traditional varieties in Latin American centers of peasant agriculture. *Bulletin of Science Technology Society* 23:350–359.

Brush, Stephen B. 1991. A farmer-based approach to conserving crop germplasm. *Economic Botany* 45:153–165.

———. 2004. *Farmers' Bounty: Locating Crop Diversity in the Contemporary World*. New Haven, CT: Yale University Press.

Brush, Stephen B., ed. 2000. *Genes in the Field: On-Farm Conservation of Crop Diversity*. Rome: International Development Research Centre (Canada); International Plant Genetic Resources Institute.

Buruchara, Robin A., L. Sperling, P. Ewell, and R. Kirkby. 2002. The role of research institutions in seed-related disaster relief: Seeds of hope experiences in Rwanda. *Disasters* 26(4):288–301.

Carrizosa, Santiago, S. B. Brush, B. D. Wright, and P. E. McGuire. 2004. Accessing Biodiversity and Sharing the Benefits: Lessons from Implementing the Convention on Biological Diversity. Environmental Policy and Law Paper 54. Gland, Switzerland: International Union for Conservation of Nature and Natural Resources.

Celis, C., M. Scurrah, S. Cowgill, S. Chumbiauca, and J. Green. 2004. Environmental biosafety and transgenic potato in a centre of diversity for this crop. *Nature* 432:222–225.

Chandler, S., and J. M. Dunwell. 2008. Gene flow, risk assessment and the environmental release of transgenic plants. *Critical Reviews in Plant Sciences* 27:25–49.

Cohen, J. I., J. T. Williams, D. L. Plucknett, and H. Shands. 1991. *Ex situ* conservation of plant genetic resources: Global development and environmental concerns. *Science* 253:866–872.

Dutfield, G. 2000. *Intellectual Property Rights, Trade and Biodiversity.* London: Earthscan.

———. 2004. *Intellectual Property, Biogenetic Resources and Traditional Knowledge.* Sterling, VA: Earthscan.

Ehrlich, P. R. 2002. Human natures, nature conservation, and environmental ethics. *BioScience* 52:31–43.

Engels, J., A. Ebert, I. Thormann, and M. Vicente. 2006. Centers of crop diversity and/or origin, genetically modified crops and implications for plant genetic resources conservation. *Genetic Resources and Crop Evolution* 53:1675–1688.

Fowler C., and P. Mooney. 1990. *Shattering: Food, Politics, and Loss of Genetic Diversity.* Tucson: University of Arizona Press.

Fowler, C., and T. Hodgkin. 2004. Plant genetic resources for food and agriculture: Assessing global availability. *Annual Review of Environment and Resources* 29:143–179.

Gonzales, Tirso A. 2000. The cultures of the seed in the Peruvian Andes. In *Genes in the Field: On-Farm Conservation of Crop Diversity*, edited by Stephen B. Brush, 193–216. Rome: International Development Research Centre (Canada); International Plant Genetic Resources Institute.

Hodgkin, Toby A., H. D. Brown, T. J. L. van Hintum, and E. A. Morales, eds. 1995. *Core Collections of Plant Genetic Resources.* Chichester, UK: John Wiley and Sons.

Hunn, Eugene. 1999. The value of subsistence for the future of the world. In *Ethnoecology: Situated Knowledge/Located Lives*, edited by V. D. Nazarea, 23–36. Tucson: University of Arizona Press.

Ingold, T. 2000. *The Perception of the Environment: Essays on Livelihood, Dwelling and Skill.* London: Routledge.

International Plant Genetic Resources Institute. 2001. Agricultural Biological Diversity. On Farm Management of Crop Genetic Diversity and the Convention on Biological Diversity's Programme of Work on Agricultural Biodiversity. Rome: International Plant Genetic Resources Institute.

Kloppenburg, Jack R. 1990. No hunting: Scientific poaching and biodiversity. *Z Magazine*, September, 104–108.

Maxted, N., B. Ford-Lloyd, and J. G. Hawkes. 1997. Complementary conservation strategies. In *Plant Genetic Conservation: The In Situ Approach*, edited by N. Maxted, B. Ford-Lloyd, and J. G. Hawkes, 15–40. London: Chapman and Hall.

Mgbeoji, Ikechi. 2006. *Global Biopiracy: Plants, Patents, and Indigenous Knowledge.* Vancouver: UBC Press.

Nazarea, V. D. 1998. *Cultural Memory and Biodiversity.* Tucson: University Arizona Press.

———. 2005. *Heirloom Seeds and Their Keepers: Marginality and Memory in the Conservation of Biological Diversity.* Tucson: University Arizona Press.

————. 2006. Local knowledge and memory in biodiversity conservation. *Annual Review of Anthropology* 35:317–335.

Nazarea-Sandoval, V. D. 1995. *Local Knowledge and Agricultural Decision Making in the Philippines: Class, Gender, and Resistance.* Ithaca, NY: Cornell University Press.

Norgaard, R. B. 1988. The rise of the global exchange economy and the loss of biological diversity. In *Biodiversity*, edited by E. O. Wilson, 206–211. Washington, DC: National Academy Press.

Pimbert, M. 1994. The need for another research paradigm. *Seedling* 11:20–32.

Plucknett, D. L., N. J. H. Smith, J. T. Williams, and N. M. Anishetty. 1987. *Gene Banks and the World's Food.* Princeton, NJ: Princeton University Press.

Prance, G. T. 1997. The conservation of botanical diversity. In *Plant Genetic Conservation: The In Situ Approach*, edited by N. Maxted, B. V. Ford-Lloyd, and J. G. Hawkes, 3–14. London: Chapman and Hall.

Rhoades, Robert E., and Virginia D. Nazarea. 1998. Local management of biodiversity in traditional agroecosystems: A neglected resource. In *Importance of Biodiversity in Agroecosystems*, edited by Wanda Collins and Calvin Qualset, 215–236. Boca Raton, FL: CRC Press.

Scurrah, M., C. Celis-Gamboa, S. Chumbiauca, A. Salas, and R. Visser. 2008. Hybridization between wild and cultivated potato species in the Peruvian Andes and biosafety implications for deployment of GM potatoes. *Euphytica* 164:881–892.

Shiva, Vandana. 1993. *Monocultures of the Mind: Perspectives on Biodiversity and Biotechnology.* London: Zed Books.

United Nations Environment Program/Convention on Biological Diversity. "On Farm Management of Crop Genetic Diversity and the Convention on Biological Diversity's Programme of Work on Agricultural Biodiversity." Synthesis paper prepared by the International Plant Genetic Resources Institute. November 5, 2001.

Vavilov, Nikolai I. 1987. *Five Continents.* Leningrad: Nauka.

Wilkes, G. 1991. In situ conservation of agricultural systems. In *Biodiversity, Culture, Conservation and Ecodevelopment*, edited by M. Oldfield and J. Alcorn, 86–101. Boulder, CO: Westview.

Zimmerer, Karl. 1996. *Changing Fortunes: Biodiversity and Peasant Livelihoods in the Peruvian Andes.* Berkeley: University of California Press.

PART ONE

Marginality and Memory in Place-Based Conservation

CHAPTER ONE

Temptation to Hope

From the "Idea" to the Milieu of Biodiversity

VIRGINIA D. NAZAREA

Portrayed as at once a crisis and a cure, biodiversity captured public imagination and became a rallying point in the 1990s. Ever since the publication of E. O. Wilson's landmark volumes *Biodiversity* (1988) and *Biodiversity II* (Reaka-Kudla et al. 1997), biological and social scientists have been analyzing causes and trends and fashioning solutions. Conservation biologists established their reputation but also risked their credibility by eschewing scientific detachment for passionate defense of biodiversity, alerting society to a growing threat of biodiversity loss fueled by population growth and increasing rates of consumption, overharvesting and habitat destruction, and introduction of exotic species (Myers 1996; Chapin et al. 2000). Somewhat overstepping their disciplinary training as biologists in favor of their moral obligation as human beings and citizens of the planet, they warned that the dominance of economic considerations over ethical-ecological ones only exacerbates these environmental and cultural vulnerabilities (Norgaard 1988; Ehrlich 2002).

Equally vocal critics have countered that since evolution is an ongoing process, with species lost and species gained at all times, biodiversity conservation is likely just an alarmist call, with shades of conspiracy to create mass hysteria, or a charismatic lure to generate more funding. Deconstructing the groundswell of scientific and public outcry regarding

19

the loss of biodiversity, these critical scholars pointed out that biodiversity is but an "idea," meaning a social/political construct and a "historically produced discourse" (Takacs 1996; Escobar 1998; Hayden 2003). Writing on *The Idea of Biodiversity: Philosophies of Paradise*, David Takacs called attention to "the making of biodiversity" and argued that "biodiversity's phenomenal success as an avatar of a scientific discipline, and a recharged conservation movement which that discipline seeks to buttress, may be its own undoing" (1996:336). From this perspective, biodiversity was cast as a dominant discourse meant to renegotiate the interconnectivity between ethnicity and territoriality, along with natural resource claims.

In order to better appreciate the ensuing dialectic, we need to understand earlier efforts, which concentrated on collecting and cataloguing biological diversity, moving germplasm from their natural habitats in centers of domestication and diversity to cold storage in designated repositories, and paving the way not only for *ex situ* conservation in genebanks but also for the formal or commercial seed industry (Plucknett et al. 1987; Kloppenburg 1988; Fowler and Mooney 1990; Fowler 1994). This was the era of the great "gene hunters" who explored distant corners of the planet and brought back prized specimens, from the little known to the sought after—a time when national sovereignty over germplasm was unheard of, and probably inconceivable. The genetic bonanza that resulted, coupled with the ominous threat of a population explosion outstripping the carrying capacity of the environment, paved the way for the Green Revolution and the breeding of modern high-yielding cultivars— the so-called miracle varieties—of the world's staple crops.

From the twin strategy of search-and-freeze and search-and-deploy, there has been a significant redirection in biodiversity conservation strategy toward more dispersed and on-the-ground efforts that acknowledge the contribution of indigenous populations, women, and small-scale farmers (Wilkes 1991; Nazarea 1998; Soleri and Smith 1999; Jarvis and Hodgkin 2000). International organizing and advocacy resulted in the Convention on Biological Diversity (CBD), signed at the Rio Earth Summit. The CBD defined biodiversity as "the variability among living organisms from all sources and the ecological complexes of which they are part" (United Nations Environmental Programme 1994:4). For the first time, national sovereignty over plant genetic resources was recognized, as the CBD bound signatory countries to "regulate and manage biological resources important for the conservation of biological diversity" and "respect, preserve, and maintain knowledge, innovations, and practices of indigenous and local communities embodying traditional lifestyles relevant to the sustainable use and conservation of biological diversity" (8–9). Despite unresolved issues pertaining to rights and responsibilities (see chapter 9 and 10, this volume), the CBD has made it difficult to ignore the crucial role of local custodians in maintaining biodiversity.

Scientists and plant genetic resources managers began to stress complementation of scientific and cultural approaches to conservation, inclusion of neglected or underutilized crops in conservation mandates, and integration of conventional functions of genebanks with new ones as dictated by evolving discoveries and political situations (Pimbert 1994; Hammer 2003). Increasingly, on-farm or *in situ* conservation, buffer zones and protected areas, and biological corridors were put in place (Cohen et al. 1991; Brush 1991; Soulé 1993; Maxted et al. 1997). Repatriation of germplasm conserved in genebanks to communities where they originated but from which they have since "disappeared" is also being explored. Overall, this new direction in plant genetic resources conservation reverses an earlier trajectory, which was to collect, systematize, sanitize, and centralize germplasm of the world's important crops in botanical gardens and genebanks (see the introductory chapter to this volume).

In a related dialectic, still in connection with biodiversity conservation but more squarely planted in the social sciences, there is an ongoing reexamination of the cultural and political dimensions of biodiversity, the *concept*, and biodiversity conservation, the *plan*. A growing body of work is examining the complex web of interactions among culture, society, and biodiversity, spurred by increasing recognition of the need for complementing formal and institutional approaches such as *ex situ* conservation in genebanks with more informal, local initiatives such as repatriation and *in situ* conservation of traditional crops (Zimmerer 1996; Nazarea 1998; Brush 2000; Cleveland and Soleri 2002). From purely ethnobotanical investigations, research in this area intensified with the negotiations for access to and benefits from plant genetic resources and associated local knowledge (Stephenson 1999; Hayden 2003; Dutfield 2004). More than two decades of ethnographic research in different regions of the world has shed light on the cultural dimensions of the nurturance and enhancement of agrobiodiversity, including local knowledge pertaining to traditional crops and varieties, and the politics of plant genetic resources conservation and use (for useful reviews, see Orlove and Brush 1996; Collins and Qualset 1998).

Yet postmodern critique questioned what it deemed to be an overly romanticized image of local environmental and agricultural knowledge and its portrayal as static and perfectly attuned to its surroundings. Earnest attempts to systematically document and bank this knowledge for biodiversity conservation specifically, and for environmental conservation in general, were viewed as misguided at best and suspect at worst (Ellen and Harris 2000; Escobar 1998; Agrawal 2002). At its extreme, critical scholarship questioned the very existence of a local reservoir of knowledge and memory, arguing that local discourse could simply be a reflection of global rhetoric and agenda. Against this intellectual conundrum, it is heartening to note that while the concept of biodiversity is being analyzed and problematized on the big stage of science, technology, and society, local farmers and gardeners are quietly celebrating and passing along their cultural and biological legacies and powerfully defying the somber politics of loss. From the privileging gaze on text and discourse, a side glance at sensory memory and quotidian practice invites us to refocus on a pivotal question that we have tended to overlook: what moves people to do what they do? It has even occasioned a temptation to hope.

From Dialectics to Dwelling

Although insignificant by themselves, cumulative acts of human defiance to forces that erode place and agency bolster biodiversity. While intellectual and policy debates may focus on loss, surrender, and abandonment associated with habitat fragmentation and agricultural development, it is important to acknowledge a powerful counter in marginal fields and uncaptured spirits. Sensuous recollection in sovereign spaces that people carve out of uniformity and predictability constantly "floods back" and replenishes this emptying engendered by modernity, the so-called erosion that applies to both cultural and genetic diversity. One example is women's intimate relationship to their seeds in Cotacachi, Ecuador, where seeds are gathered in the aprons of their skirts (see cover image) and transported in the folds of their clothes, to be hidden and protected or displayed and shared, as they see fit (Nazarea et al. 2006:8). In Cusco, Peru, farmers bundle up their newly harvested potatoes and sing to them, cradling them as their *wawas* (infants). Errant varieties that dare to "disappear" temporarily are scolded when they are found again. Andean farmers "know" their crops by remembering. They "walk" their seeds, displaying and exchanging them to enrich and revitalize their germplasm (Valladolid and Apfel-Marglin 2001). For them, biodiversity is life itself and not to be treated as a good or ward, as an entity separate from themselves.

Cultural memory embedded in food and place enables small-scale farmers and gardeners to resist wholesale surrender to what Vandana Shiva (1993) calls "monocultures of the mind." Even under serious threat of diminishment, they continue to cultivate a wide array of plants, sustained by recollections regarding the plants' aesthetic, culinary, and healing qualities, as well as their ritual significance and connection to the past (Richards 1986, 1996; Nabhan 1989; Nazarea 1998, 2005). Here, I present stories of resistance, resilience, and recovery in relation to "lost" and "found" potatoes. My intention in highlighting these *small movements* is to examine a kind conservation that rests on memory and sense of place and to explain why these elements are so central to the nurturance of biodiversity at different levels. As I will try to demonstrate, these senses of place serve as tenacious footholds for resisting the obliteration of agency and choice. This resistance to homogeneity on which the maintenance of diversity depends is based, literally and figuratively, on "re-remembering" and digging in.

According to Edward Casey, sense of place "imbues a coherence at the primary level, one supplied by the horizons and depth of experience" (1996:18). People have to trust this coherence and integrity or they cannot function. Grinding against and sifting through everyday experience, local knowledge and cultural memory are always in the process of construction and repair, to such an extent that some authors have spoken of the "afterwardness" of memory (Tonkin 1992; Haug 2000; Rigney 2005). But a sense of place is like a thread that runs through a fragile warp, one that makes hope of "getting back in place" possible (see chapters 11 and 12, this volume). Beyond the purely cognitive, or the purely rational, a "feeling" for one's surroundings comes from an intensive apprenticeship or "enskillment" in its idiosyncrasies and demands and endows one with a habit of mind and a bodily orientation that is honed in place (Merleau-Ponty 1962; Ingold 2000, 2005). Subconscious or pre-attentive affective frameworks emanate from one's locality, giving rise to "dispositions in positions" (Bourdieu 1987). These predispositions constitute our lenses and thereby inform practice; thus, they define latitudes of choice in our decision making and behavior.

Paradoxically, both science and its various critiques disempower place and agency in its treatment of local knowledge and memory in relation to biodiversity. We tend to ignore or downplay the sensory embodiment—remembered and longed-for aromas, flavors, and compositions—as well as the attendant emotion that make people hold on to "forbidden," "quirky," and "archaic" plants, no matter how unwise, unprofitable, or inconvenient. Performed and recreated in everyday practice, neither place nor agency is logically formulated or inscribed as a set of processes or principles. There is a need to reinsert embodiment into the conversation about conservation, and to materialize, in "concrete utopias" (Bloch 1988) that dot the landscape of forgetting, a vision of sensuous dwelling that cannot be subject to a universal standard or programmatic design. As Paul Stoller notes, "Embodiment is not primarily textual; rather, the sentient body is culturally consumed by a world filled with forces, smells, textures, sights, sounds and tastes, all of which trigger cultural memory" (1994:636).

Milieu of Memory

A hunger for memory and connection seems to possess the present time, characterized as it is by displacement, alienation, and homogeneity. While memory is not history, a new scholarly interest predicated on longing to recall, if not to relive, the past blossomed, ironically, in history. Mourning the loss of the milieu of memory as represented by rural life before the advent of modernity, Nora called attention to

> an increasingly rapid slippage of the present into a historical past that is gone for good, a general perception that anything and everything may disappear . . . a rupture of equilibrium. The remnants of experience still lived in the warmth of tradition, in the silence of custom, in the repetition of the ancestral, have been displaced under the pressure of a fundamentally historical sensibility. We speak so much of memory because there is so little of it left. . . . Indeed, we have seen the tremendous dilation of our very mode of historical perception, which, with the help of the media, has substituted for a memory entwined in the intimacy of a collective heritage the ephemeral film of current events. (1989:363)

Noting that "the real environments of memory are gone" but that present-day memory "crystallizes and secretes itself in the *lieux de mèmoire,* or sites of memory," Nora undertook an ambitious project to document France's rich cultural legacy. In three heavy tomes dedicated to *Realms of Memory* (originally published in French from 1968 to 1992) Nora acknowledged that "the bed of memory cannot be extended indefinitely" (1992:367) and lamented that "modern day memory is archival memory."

But is the milieu of memory gone for good, "lost" like native or rustic potatoes? Or is there a hidden realm where memory is nurtured and lived? In secret recesses close to the heart, such as small private orchards and backyard gardens, agency and hope emanate from invisible threads of connection to a shared past. Memory is "placed" or embodied in actions and objects, incorporating values that are smuggled into the present and the future. Though mostly shredded and patched up, these memories are nonetheless cherished, even in exile. Thus, Vietnamese immigrants in the United States layer sweet potatoes, lemon grass, bitter melon, and banana plants in furrows and trellises in their yards and savor steaming bowls of aromatic herb-flavored noodle soup and fresh hand-rolled spring rolls in their kitchens to summon an "out of place sense of place" (Nazarea 2005). Urban professionals while they were in Vietnam, many political refugees in the United States have taken up vegetable gardening for the first time in their lives, growing plants from the homeland in their adopted backyards—and, sometimes, to the dismay of some of their neighbors, in their front yards—to anchor this authored sense of place.

The alternative scenario, loss of memory and diversity concomitant with agricultural development, tends to divest producers and consumers alike of viable options and meaningful choice. Asian rice farmers, for example, remark that if they alone planted the traditional, aromatic rice varieties, the fragrant ripening grains will attract all the mice and birds to their fields, thus drastically reducing their yield vis-à-vis other farmers. On top of this, their traditional varieties will take longer to mature, preventing them from coordinating with their neighbors and taking advantage of economies of scale by pooling resources for spraying over a wider area or hiring a tractor for cultivating their small but contiguous farms (Scott 1985; Nazarea-Sandoval 1995). For the most part, agricultural modernization has been a story of colonization—a narrowing of not only the genetic but also the cultural base of farming—a Trojan horse that keeps on giving. In its wake are cherished memories of a once-fecund milieu: following the plow in irrigated rice fields to catch edible fish and frogs, resting between planting and harvesting low-subsidy crops, cooking porridges and sweets of various textures and flavors, exchanging seeds and slips, and celebrating the rhythms and seasons of life through ritual and commensality.

It is not surprising that while the history of the Green Revolution is replete with success stories of rising productivity and income with greater use of modern inputs, the memory of the Green Revolution is something else: of women working long, tedious hours; of farmers driven to debt, deformity, derangement, and suicide; of land stripped and drained due to highly extractive double- and triple-cropping regimes. In the frontier town of Nanegal, Ecuador, Don José characterized his situation as being *fregado* (washed out, without hope): "They now put three *quintales* (one quintal is approximately equivalent to 100 pounds) of fertilizer for one quintal of potatoes so, sure, if you see a small piece of land there is no space to step between the plants. It is full of plants because they give them pure ingredients, nutrients for the flowers, for the roots, and so on" (Nazarea 2005:210–211). Meanwhile, in highland Cotacachi, Magdalena Fueres cultivates medicinal herbs, flowering plants, and fruit trees in her homegarden (see figure 1.1). Magdalena was a full-time housewife in her hometown of La Calera before she became a dedicated leader of *Jambi Mascaric* (Search for Health) charged with coordinating community outreach aimed at complementing conventional

FIGURE 1.1. Magdalena Fueres in her homegarden showing us her tomate de arbol. Photo by Robert E. Rhoades.

Western medicine with indigenous health practices. She has been extremely successful in promoting maternal and child health and bilingual literacy, always prioritizing local resources and needs and refusing paternalism and other forms of dependence (see the introductory chapter and chapter 5, this volume). Magdalena named her first and second sons, who were born before the indigenous resurgence, Christian and Darwin. Then she named her youngest, a daughter, Mishary, which in Quichua means "triumph" or "victory." In *Recipes for Life*, Magdalena conveyed to her children "how important the wisdom of our elders is yet we do not value it because we have been led to believe that those who don't read and write don't have knowledge. Now I know how important the wisdom and practices of our people are, but we are forgetting that knowledge" (Nazarea et al. 2006).

In Cusco, Peru, efforts are under way to return to Quechua[1] farmers hundreds of potato landraces that have been collected over the years and stored at the International Potato Center (Centro Internacional de la Papa, CIP) genebank in Lima. Six interlocking communities have constituted themselves into the Parque de la Papa (Potato Park) and are reintegrating these native potatoes into local farming systems. Complemented by other initiatives such as documentation of customary laws and revival of culinary traditions in the Andes, the process of repatriation in Cusco does not stop with potatoes, or with projects, but stirs interest and awakens memory in multiple aspects of place and agency (see figure 1.2). Farmers are not only retrieving "lost" tubers but also revitalizing seeds of wisdom and resistance. The following exchange between two women who participated in our initial meeting with Quechua communities at the Potato Park hints at the cultural reinvigoration that is taking place:

> YOUNGER WOMAN: We use wild greens that have to go with the potato. We cannot find them; now we have to walk long distances to find them. We used to have different crops. We used grains, beans, and other things, but we also can't deny that we have been influenced and we are using noodles. Sometimes what I do is combine those noodles with what we have.
>
> OLDER WOMAN: I cook everything. I find everything. I know where to find the wild things. I know where they are and then you have to use them. You don't have to deplete them. These aromatic herbs, you have to know how to use them so they can be tasty. Those herbs are necessary for making good food. I cannot forget that. Even

FIGURE 1.2. Young Quechua women trained to direct the lens in documenting the repatriation project in Cusco, Peru. Photo by Robert E. Rhoades.

when I'm dead, I'm not going to forget it! (Laughter) I'm teaching
my children and nieces and nephews about old foods so they can eat
well. I don't like eating those lowland foods like rice and noodles.

According to Pierre Nora, "If we could live in memory, we would not
have to consecrate sites of memory to its name" (1997:366). The living
milieu is where the past continues to be engaged through everyday activities like gardening, cooking, sharing, and celebrating. What is pecu-
liar about sensory memories in connection with biodiversity conservation
is how they make alternatives prominent and compelling to the extent
that it is difficult to tell if we are dealing with more discrete and, in a
sense, more intentional "sites of memory" or with the throbbing "milieu
of memory" itself. With modernity's dislocations and its discontents, often the lines are blurred, and in some cases it is necessary to start with a
dedicated "site" like the Potato Park to peel back the layers of forgetting
and thereby reawaken the "milieu." In the end, where it begins and
where it ends is of secondary importance to the restoration of the full-
ness, complexity, and depth of life that only the conservation of diversity
can sustain.

Landscape of Resistance

Why people hope enough to resist, and why they resist enough to persist,
depends, in large measure, on the past that is in their present. From an
inspired quest for "history from below" and "history of everyday life," the
study of memory became the window for understanding human engagement with the past, with landscapes, with food, and with biodiversity
(Nazarea 1998, 2005; Sutton 1998, 2001; Harkin 2003; Gordillo 2004).
While this engagement may be at a very personal level, agency and resistance are sustained by memory that is social, collective, and lived. People
acquire or construct memory in communion with kin and peers, so
much so that the analytical distinction between individual memory and
social memory is meaningless (Halbwachs 1992). Paul Connerton sug-
gested that, "if we are to say that a social group, whose duration exceeds
that of the lifespan of any single individual, is able to 'remember' in common, it is not sufficient that the various members who compose that
group at any given time should be able to retain the mental representation relating to the past of that group. It is necessary also that the older
members of that group should not neglect to transmit these representations to the younger members of the group" (1989:38).

Cultural memory is vital for "people without history" (Wolf 1982).
People who were colonized and converted, displaced and disinherited,

and in other ways marginalized elaborate an alternative version of the past that is counterhegemonic and subversive—and ultimately healing— constituting a potent *countermemory* (Boddy 1989; Lipsitz 1991; Zerubavel 1995). Trauma in relation to memory is a potent framework from which to view environmental changes and development schemes that destroy whole landscapes rich in biological and cultural legacies (Steinberg and Taylor 2003; McDowell 2004; Cruikshank 2005; Nazarea 2005). More specifically, in relation to genetic erosion, trauma comes from the aggressive promotion of modern varieties and imported technologies to the extent that all choices in crops and their cultivation are "disappeared" (Shiva 1993; Dove 1999). It also results from the assumption that science and its products are neutral, as in the pursuit of culturally uninformed or politically motivated plant breeding and biotechnology. When watery, tasteless sweet potatoes are substituted for dense, delicious varieties, or when herbs used in everyday practice and ritual are "genericized" and commercialized with neither an appreciation of the social context of their use nor prior informed consent from their custodians, violence is perpetuated on them and all humanity. Such intrusions banish attachment to place and anticipation of tastes and seasons, thereby engineering an "epoch of tastelessness" (Seremetakis 1994).

In the repatriation project at the Potato Park, people are digging in and reseeding the "bed of memory" (Nora 1998). They are working through hegemonic models imposed over indigenous cosmologies and slowly dismantling the economics of detachment that has been layered over the intimacies of place. Over four hundred potato landraces have been brought back to Cusco from the genebank in Lima and a historic Repatriation Agreement has been forged between CIP and the Quechua farmers from the six participating communities (see chapter 10, this volume). ANDES,[2] a Peruvian nongovernmental organization (NGO), brokered the agreement—a much-lauded first in the international conservation arena that unequivocally recognizes the contribution of farmers to crop genetic diversity and compensates them for the benefits derived from its use. Needless to say, the Potato Park is attracting the kind of international attention that only the restoration of "lost" crops in the navel of Inca civilization can.

But it should be noted that the CIP collection did not accrue in a vacuum, the advocacy and action did not foment out of the blue, and the potato landraces are not being reinstituted in fields devoid of diversity. Although communal welfare is extolled over individual accomplishment in Quechua communities, the most highly respected farmers are still those individuals who cultivate the greatest variety of potatoes. Women are admired for their skill in the selection and preparation of richly varied and finely suited potatoes for offering, planting, processing, and eating.

There have always been cultural practices for enhancing and refurbishing this diversity, and there are Quechua songs and rituals both for calling back and letting go of spirits, lovers, birds, and potatoes. We should also not lose sight of the fact that there were scientists who, for sentimental reasons or national pride, searched for and conserved these potatoes, as well as intellectuals and activists who advocated for their return and reintegration into Quechua fields. Between the loss and the recuperation are men and women who held on, and held fast, to diversity within the milieu of memory.

Insubordinate Lives

Impressed by the repatriation of potato landraces to their source communities, and their reincorporation into indigenous farming systems, we sought out and interviewed CIP scientists, NGO workers or *técnicos*, and potato farmers or *paperos*.[3] The interviews were in the form of a more or less unstructured or free-flowing life history elicitation conducted in English, Spanish, and Quechua, respectively. Interviews were recorded, transcribed, and translated by competent and, whenever possible, native speakers. To examine threads of resistance to the erosion of cultural and biological diversity, I have selected three life histories, those of (1) Carlos Ochoa, a Peruvian plant collector and breeder who taught at the Agricultural University in Lima and then moved to CIP shortly after it was established; (2) Orestes Castaneda, a Quechua papero and a local ANDES técnico who grew up in Sacaca, one of the communities comprising the Potato Park; and (3) Luisa Huaman, a Quechua farmer and grandmother who remembers a time before eucalyptus, noodles, and aluminum pots.

Carlos Ochoa, a world-renowned scientist who has collected more wild potatoes than any other plant explorer, put himself through school in Bolivia—studying by day and working as a math teacher by night—even though he came from a wealthy family. He remarked that life was "a little hard" but wanted to keep the reasons private. Upon his return to Peru, his father asked him to manage the family business, but he decided instead to work at an experiment station in the Mantaro Valley. After working there for some time, he thought, "Five years as a wheat breeder, this is a stupid thing for myself. Never was I really a good wheat breeder. Wheat never really did a good thing for this country. . . . It's a country for potatoes and, against my own chiefs, I started to do the first work in potato breeding." He bred two successful varieties in the 1950s, one he named Mantaro after his first research site and the other *Renacimiento*, meaning "rebirth." Unsatisfied with his job, and wanting to teach again, he moved to the Agrarian University at La Molina. There he founded the National Potato Program, which exists to this day. He proposed that the program be elevated to a national institute, and the proposal was so well received that he thought it might actually become a reality. But at around the same time the North Carolina Agricultural Mission gained the support of the Rockefeller Foundation and laid the groundwork for CIP in Lima.

Recruited to the newly established center with offers of superior working conditions and research support that his university could not hope to match, he moved to CIP. There he was tasked with forming the world germplasm bank for potatoes, a demanding responsibility that precluded breeding. This was not something he had anticipated, since breeding potatoes was his real love. For him, "it was very, very hard," but there was no turning back because he had already given up his post at the university. So he put his energy into collecting wild potatoes throughout the Andean region and appreciated the facilities, travel opportunities, and scientific connections that only an international center like CIP could offer. Trekking through mountains and valleys all over South America and exploring every possible niche, he discovered more than eighty species, close to half of all wild potatoes known to science. And yet, passionately, he bred potatoes on the side.

Contrary to the prevailing orthodoxy that emphasized a strict and narrow breeding design, Dr. Ochoa's strategy was guided primarily by local concerns and preferences. He started from varieties farmers recognized, "those that they had feelings for." Using *papas rusticas*, or rustic potatoes, as progenitors instead of official breeding lines, he aimed his breeding at finding solutions to problems that mattered and came up with varieties resistant to late blight, which posed the greatest threat to farmers. He also obsessed about acceptability and taste. His Renacimiento was "good for making money" but did not quite match the traditional favorite in Cusco called *qompis*, which was much desired by farmers and consumers alike because it was large and delicious. According to Dr. Ochoa, the comparison between the two haunted him, "like music to my ears." With qompis as his muse, he worked hard for seventeen years until he bred "a good one" that he named Micaela Bastidas after the wife of the Incan rebel Tupac Amaru, who was executed by the Spaniards in the square as she watched. He produced another popular variety that he named Tomasa after the lieutenant of Tupac Amaru who died tragically in a place that became his family's farm in Cusco. According to some accounts, Tomasa was also the name of his beloved nanny.

All of Dr. Ochoa's varieties were named to recognize and honor special people, places, and events that held meaning in his heart. A sentimental man, he has cried for place for much of his life; first as a struggling college student in Bolivia and later as a graduate student in Minnesota, USA, where it was not the cold that bothered him so much as the solitude: "Lonely, lonely, lonely . . . lack of affection of family, of friends, I suffered terribly." He cared most for "what's inside you . . . that which you can't turn away from," and this passion invariably guided his life and his career. Eventually, he was reprimanded for his breeding work and accused of dishonoring his *compromiso* (gentlemen's agreement). Now, having published several authoritative volumes on taxonomy and ethnobotany and received international recognition for his work, he reflects back on his clandestine breeding activities: "Maybe I did (violate the agreement), but I thought I was doing a nice thing, because I like to do that. Up to now, I like to." Ironically, Dr. Ochoa's varieties, bred to address particularities of locality and matters of the heart, were the ones accepted throughout the nation and the world, bringing him quiet pride.

Orestes Castaneda remembered his father's *chacras* (fields) that yielded abundant food. His father worked hard; accordingly, "we only bought matches to cook, kerosene to light the house, salt, and pepper. And because we had livestock, we dried meat and used it little by little. . . . We did not lack cheese, milk; because there were a lot of products, we nourished ourselves well." They cooked in open hearths with clay pots and wooden spoons, unlike the aluminum and plastic ones they used today. He reminisced "in clay pots you can feel the flavor of the produce and they tasted better; in clay pots, food does not get cold, it stays warm." There was abundant *chuno* and *moraya* (two kinds of indigenous freeze-dried potatoes), and, according to Orestes, "in the storage room we stored chuno, favas, wheat, potatoes. . . . Now, rice, noodles, preserves, and dried fish are easily found (but) food is no longer natural, it is artificial, even tasteless; it can also cause discomfort."

For Orestes, there are only two kinds of changes—the good changes and the bad ones. Among the good, he counts schooling for children, cohesion among communities, and health clinics for everyone. There is a downside to each of these, though, that makes Orestes nostalgic for the past. Schools give children opportunities that their parents never had. And yet because of schooling, children shun their parents and their traditional ways and no longer want to work hard in the chacras. He complained that "before, people respected the land and they made *pagos* (offerings) to the *apus* (gods). Today there is not even respect for plants." While people in their communities are quicker to patch up their differences and better organized to deal with conflicts, Orestes still feels that in the time of his grandparents there was more harmony and love, and certainly respect. Today, children do not want to be seen with their elders and elders are insulted when young ones greet them in casual citified ways. Health posts, by dispensing pills and advice, benefit the people but also promote beliefs and practices that make them weaker. He observed the declining resistance of children, explaining that "we did not grow up all wrapped up and yet we were healthy; now we wrap our children up so they won't get sick and in spite of it they get sick." He also noted that they used to heal themselves with herbs. With the coming of the clinics, first the knowledge about herbs, and then the herbs, disappeared.

Kausayniykuy, a Quechua concept for well-being, means, for Orestes, all the necessities and comforts they possess like their house, their chacra, their furniture, and their clothes. It also encompasses the rocks, the lakes, the wild animals, the trees, and the herbs around them. He places community well-being above individual well-being and admonishes that "we should take care of everything around us." Most of the bad changes involve a deterioration of these elements. Climate has become extreme and unpredictable; rain does not come when you expect it, resulting in draught followed by strong rains and heavy frost that kill their crops. Natural springs are drying up, leading to water scarcity, and eucalyptus trees make matters worse by draining water and nutrients from the soil. Herds of cattle, sheep, and even *cuys* (guinea pigs) are decimated with diseases. As a child he ate "natural meat" but now meat is "poisoned." They used manure and other organic matter to enrich the soil for growing potatoes when he was growing up and nothing more. He can still recall when improved potatoes appeared and fertilizers were first introduced.

The point of rupture is marked by "potatoes that are neither good nor tasty," potatoes that end up small, contaminated, and green (poisonous). "Worms" is another marker; it stands for a host of diseases that now plague crops and livestock. He was reminded of how delicious the meat he ate as a child was when he visited a more remote community that did not have slugs and they served him good, tasty meat. As children, they followed the herds and watched them graze. They could lie down on the hills for a long, long time because the sun did not burn. Their parents used herbs to nourish and heal their children. They also told them stories like the one about the soup potato, *leq'echu*, that came from the apus Pitusiray and Sawasiray: "The soul of this little potato had been transformed into a bird and there was an elderly man who saw this bird sitting over a rock and crying. When he came close, there was no bird but a potato was in its place. Since then he began to cultivate this potato leq'echu." Another story they told was about San Gregorio who worked well with potatoes. When God's emissary approached him and asked what he was planting, San Gregorio answered "qhullurumita." This was a kind of small potato. Had he answered "sunturuma," a kind of big potato, then potatoes would have been big in their communities and elsewhere.

Orestes is pleased that, with the help of ANDES, their communities are beginning to gain national and international recognition. Through his work as an ANDES técnico, he met people coming from faraway places to witness their agricultural practices and rituals who left with an appreciation of the need for repatriation and *in situ* conservation. He considers the Potato Park as their unique contribution since in other parts of the world, "there are only parks with animals, mountains, and forests but there is none with agrobiodiversity." He is also happy that nowadays the communities have ceremonies in honor of San Gregorio, Pitusiray, and Sawasiray. Orestes revealed that "just now we are beginning to value the teachings of our forefathers and just now we are noticing how important it is to maintain our knowledge. . . . We always tell our sons about the chacras, about the potatoes. Our daughters stay more with their mothers who teach them the name of the apus, how to cook, to wash, the chores with herds, to cure with herbs. We always tell."

Luisa Huaman is referred to in her community as *anciana* (elderly woman) and deferentially addressed as *Mamita* or Grandmother. She lives on her own but close to her daughter and son-in-law in the community of Cuyo Grande. At the time of the interview she was helping her daughter to cook breakfast, preparing a drink made with toasted corn and ground fava and a potato soup. She recalled that in her time they planted potatoes in containers called *puqos*. In the hole they would put guano, three or four small potatoes and, in the middle, three coca leaves as an offering to the apus. According to her, "before planting the potato one must make a payment and then you can blow some alcohol or *chicha* (a fermented drink) to the apus." She remembers that in her youth "everything was silent"; there were no cars, not even trees, just grasses and spiny bushes like *kantukiswar*, *quena*, and *chachacomo*. They gathered spines and used these as fuel for cooking. There were no eucalyptus trees.

In her youth, they planted legumes like fava, grains like wheat and barley, and tubers like *olluco*, *oca*, and *anu*. They had a lot of chickens and cows and brought eggs and cheese to the city to sell. They also raised pigs and so there was a lot of lard. They used lard to flavor their dishes along with *yuyu* or wild greens that they would collect. As Luisa recalled:

> Before, we ate well because products were large and tasty. The potatoes were big and there were many different varieties like *pasnacha*, *peruntus*, *serqa*, *qompis*, *pukanawi*, *churos*, *pukachuros*, and *chillico*. The pasnacha potato was pretty, rich-tasting, and yellow. . . . Before, one would kneel before the apus; now people don't feel like making payments because production is no good. In the past, when you ran in

the chacras the potatoes would come out of the soil and it would not be necessary to dig. The fava beans were also large and tasty. I would like to go back to that time.

At present, they no longer plant *papa huayco* "because native potatoes get worms." Now they only plant *papa chaska*; now potatoes "are very small and turn green." They are not mealy in texture like before. Mamita Luisa complains when her daughter cooks store-bought noodles or rice for her family. While growing up, she was nourished with hearty barley soup, wheat soup, and corn soup and so she refuses to eat things she was not familiar with as a child. She also remembers cooking only in clay pots that they had to go to San Salvador to purchase. According to her, "There were no aluminum pots, even high-class ladies used clay pots." Like other women in the community, she cooks in whatever vessel is available, even though she knows that anything cooked in a clay pot always comes out richer in flavor. But, for herself and her loved ones, she insists on preparing invigorating drinks and tasty soups that she knows from memory.

Sensuous Conservation

A continuum of countermemories, from those inscribed in subversive discourse to those incorporated in performative ceremonies and sedimented in daily practice, serves as an alternative rendering of history, correcting an imbalance and creating a powerful consciousness (Rappaport 1998; Stoller 1995; Lambek 2002). Of particular relevance to biodiversity are sensory memories of places and food, of garden plots seeded with living, edible legacies, meandering trails where pungent, healing greens are gathered, and simple meals that evoke connections to the past as well as to family and home. Like fairytales that "move on their own in time," these memories help float and eventually congeal a "wish landscape" over a constraining one (Bloch 1988; see also Nazarea 2005). Thus embodied, countermemory enlivens agency and presents choices emanating from the availability and accessibility of a diversity of plants and the complex lifeworlds that they help imagine and make tangible.

Since there are obviously some very practical limitations to complete and constant dwelling in the milieu of memory, perhaps the best we can hope for is a melding of the milieu and its various sites in the spirit of sensuous conservation. In the case of the three Peruvians whose lives, longings, and concerns I have attempted to capture in the preceding section, we can see that there is no unadulterated dwelling in place or steadfast clinging to the past. There is, instead, the trauma of loss replenished by embodied memories of what they have found useful, tasty, beautiful, moving, or meaningful. Carlos Ochoa lived and studied in the United States and published widely according to Western scientific modes and standards; Orestes Castaneda imbibed conservation rhetoric from his exposure to ANDES and often used it as a framework for assessing the past as well as the changes and prospects around him; and Luisa Huaman has acquiesced to preparing traditional potato soup using aluminum pots. By turns they have forgotten and remembered, released and retained, thereby crafting a workable hybridity. They have anointed domains of their lives to which they keep returning, and marked off personal sanctuaries within which they remain sovereign.

Sensuous conservation sheltered in marginal spaces can effectively dismantle the "organized forgetting" or "forced amnesia" that is both the *modus operandi* and the worst possible consequence of any totalitarianism (Connerton 1989). For while historical and political forces are of crucial—one might say crushing—significance, the power of imagination and memory in crafting insubordinate lives cannot be underestimated. This suggests a new direction in conservation that would draw on emotional attachment to plants, animals, and a host of other sentient beings in the world. Reinscription and reembodiment of cultural memory and biological diversity can come in the form of "concrete utopias" (Bloch 1988)—such as Magdalena's homegarden, Mamita Luisa's cooking, or the Potato Park's fields and repositories—where landscapes of longing and hope are mapped onto constrained exterior landscapes through objects and habits that stimulate sensory recollection and affective engagement. A contagion of emotion can bring about restorative social movements to communities that have suffered from the trauma of loss of identity, agency, and diversity.

The ultimate measure of success for a repatriation initiative like the Potato Park in Cusco is not the number of potatoes brought back, or the number of customary laws and traditional dishes resurrected, but rather the reattachment of stories and memories to these potatoes, laws, and dishes such that there is little chance that they can ever again be abandoned, or taken away. The milieu of memory is alive although it is overlaid with the debris of accelerated and somewhat mindless modernization. Being hidden has its advantages, however, for marginality and illegibility are its best protection and defense. So *keeping the milieu* is a balancing act or, one might even suggest, a deliciously dangerous game. Sensuous conservation is just as confounding. But it is an open road, and inviting. "A little hard," Dr. Ochoa would say, but not impossible. What is it that sustains people? What is it that moves them? In conservation of biodiversity, the answer seems to lie in the enduring bonds to place and the past that inspire insubordinate lives.

Notes

1. The indigenous group in Peru is referred to as "Quechua" after the language of the Inca kingdom. In Ecuador, which came under Inca rule in the fifteenth century, the indigenous group and their language are referred to as "Quichua."

2. ANDES is an acronym for Associacion para la Naturaleza y el Desarollo Sostenible (Association for Natural Resources and Sustainable Development). A Peruvian NGO organized in 1995, it played a central role in founding the Parque de la Papa and negotiating the Repatriation Agreement treaty (also known as the Agreement) for repatriation of potato landraces that was signed in 2004 and renewed in 2010.

3. Alejandro Argumedo, director of ANDES, invited us to the Potato Park after he participated in the conference Seeds of Resistance/Seeds of Hope: Repatriation and In Situ Conservation of Traditional Crops, which led to this volume. He hosted our first meeting at the Potato Park in 2005 and many of our subsequent trips. Robert E. Rhoades, Magdalena Fueres, Juana Camacho, and Jenna E. Andrews-Swann have collaborated with me in various stages of this research.

References

Agrawal, A. 2002. Indigenous knowledge and the politics of classification. *International Social Sciences Journal* 173:287–297.

Bloch, E. 1988 [1930]. The fairy tale moves on its own in time. Reprinted in *The Utopian Function of Art and Literature: Selected Essays*, edited by J. Zipes, 163–166. Cambridge, MA: MIT Press.

Boddy, J. 1989. *Wombs and Alien Spirits: Women, Men, and the Zar Cult in Northern Sudan*. Madison: University of Wisconsin Press.

Bourdieu, P. 1987. What makes a social class? On the theoretical and practical existence of groups. *Berkeley Journal of Sociology: A Critical Review* 31:1–18.

Brush, Stephen B. 1991. A farmer-based approach to conserving crop germplasm. *Economic Botany* 45:153–165.

———, ed. 2000. *Genes in the Field: On-Farm Conservation of Crop Diversity*. Rome: International Development Research Centre (Canada); International Plant Genetic Resources Institute.

Casey, E. S. 1996. How to get from space to place in a fairly short stretch of time: Phenomenological prolegomena. In *Senses of Place*, edited by Steven Feld and Keith H. Basso, 1:13–52. Santa Fe: School of American Research Press.

Chapin, F. S., E. S. Osvaldo, C. B. Ingrid. 2000. Consequences of changing biodiversity. *Nature* 405:234–242.

Cleveland D. A., and D. Soleri, eds. 2002. *Farmers, Scientists and Plant Breeding: Integrating Knowledge and Practice*. New York: CABI.

Cohen, J. I., J. T. Williams, D. L. Plucknett, and H. Shands. 1991. *Ex situ* conservation of plant genetic resources: Global development and environmental concerns. *Science* 253:866–872.

Collins, W. W., and C. O. Qualset, eds. 1998. *Biodiversity in Agroecosystems.* Washington, DC: CRC Press.

Connerton, P. 1989. *How Societies Remember.* London: Cambridge University Press.

Cruikshank, J. 2005. *Do Glaciers Listen? Local Knowledge, Colonial Encounters, and Social Imagination.* Vancouver: University of British Columbia Press.

Dove, M. 1999. The agronomy of memory and the memory of agronomy: Ritual conservation of archaic cultigens in contemporary farming. In *Ethnoecology: Situated Knowledge/Located Lives,* edited by V. D. Nazarea, 45–69. Tucson: University of Arizona Press.

Dutfield, G. 2004. *Intellectual Property, Biogenetic Resources and Traditional Knowledge.* London: Earthscan.

Ehrlich, P. R. 2002. Human Natures, Nature Conservation, and Environmental Ethics. *BioScience* 52:31–43.

Ellen, R., and H. Harris. 2000. Introduction. In *Indigenous Environmental Knowledge and Its Transformations: Critical Anthropological Perspectives,* edited by R. Ellen, P. Parkes, and A. Bicker, 1–33. Amsterdam: Harwood.

Escobar, A. 1998. Whose knowledge? Whose nature? Biodiversity, conservation, and the political ecology of social movements. *Journal of Political Ecology* 5:54–82.

Fowler, C. 1994. *Unnatural Selection: Technology, Politics, and Plant Evolution.* International Studies in Global Change 6. Switzerland: Gordon and Breach.

Fowler, C., and P. Mooney. 1990. *Shattering: Food, Politics, and Loss of Genetic Diversity.* Tucson: University of Arizona Press.

Gordillo, G. R. 2004. *Landscape of Devils: Tensions of Place and Memory in the Argentinean Chaco.* Durham, NC: Duke University Press.

Halbwachs, M. 1992. *On Collective Memory.* Edited and translated by L. A. Coser. Chicago: University of Chicago Press.

Hammer, K. 2003. A paradigm shift in the discipline of plant genetic resources. *Genetic Resource and Crop Evolution* 50:3–10.

Harkin, M. E. 2003. Feeling and thinking in memory and forgetting: Toward an ethnohistory of the emotions. *Ethnohistory* 50(2):261–284.

Haug, F. 2000. Memory work: The key to women's anxiety. In *Memory and Methodology,* edited by S. Radstone, 155–178. New York: Berg.

Hayden, C. 2003. *When Nature Goes Public: The Making and Unmaking of Bioprospecting in Mexico.* Princeton, NJ: Princeton University Press.

Ingold, Tim. 2000. *The Perception of the Environment. Essays on Livelihood, Dwelling and Skill.* London: Routledge.

———. 2005. Epilogue: Towards a politics of dwelling. *Conservation and Society* 3(2):501–508.

Jarvis, D., and T. Hodgkin. 2000. Farmer decision making and genetic diversity: Linking multidisciplinary research to implementation on-farm. In *Genes in the Field: On-Farm Conservation of Crop Diversity*, edited by Stephen B. Brush, 261–278. Rome: International Development Research Centre (Canada); International Plant Genetic Resources Institute.

Kloppenburg, Jack R. 1988. *First the Seed: Political Economy of Plant Biotechnology 1942–2000*. Cambridge: Cambridge University Press.

Lambek, M. J. 2002. *The Weight of the Past: Living with History in Mahajanga, Madagascar*. New York: Palgrave Macmillan.

Lipsitz, G. 1991. *Time Passages*. Minneapolis: University of Minnesota Press.

Maxted, N., B. Ford-Lloyd, and J. G. Hawkes. 1997. Complementary conservation strategies. In *Plant Genetic Conservation: The In Situ Approach*, edited by N. Maxted, B. Ford-Lloyd, and J. G. Hawkes, 15–40. London: Chapman and Hall.

McDowell, L. 2004. Cultural memory, gender and age: Young Latvian women's narrative memories of war-time Europe, 1944–1947. *Journal of Historical Geography* 30:701–728.

Merleau-Ponty, M. 1962. *Phenomenology of Perception*. Translated by C. Smith. London: Routledge and Kegan Paul.

Myers, N. 1996. The biodiversity crisis and the future of evolution. *Environmentalist* 16:37–47.

Nabhan, G. 1989. *Enduring Seeds: Native American Agriculture and Wild Plant Conservation*. San Francisco: North Point Press.

Nazarea, V. D. 1998. *Cultural Memory and Biodiversity*. Tucson: University of Arizona Press.

———. 2005. *Heirloom Seeds and Their Keepers: Marginality and Memory in the Conservation of Biological Diversity*. Tucson: University of Arizona Press.

Nazarea, Virginia D., Juana Camacho, and Natalia Parra, eds. 2006. *Recipes for Life: Counsel, Customs, and Cuisine from the Andean Hearths*. Quito: Ediciones ABYA-YALA.

Nazarea-Sandoval, V. D. 1995. *Local Knowledge and Agricultural Decision Making in the Philippines: Class, Gender, and Resistance*. Ithaca, NY: Cornell University Press.

Nora, Pierre. 1989. Between memory and history. *Les lieux de mèmoire. Repre-sentations* 26:7–24.

———. 1996–98. *Realms of Memory: The Construction of the French Past*. Translated by Arthur Goldhammer. 3 vols. New York: Columbia University Press.

Norgaard, R. B. 1988. The rise of the global exchange economy and the loss of biological diversity. In *Biodiversity*, edited by E. O. Wilson, 206–211. Washington, DC: National Academy Press.

Orlove, B. S., and S. B. Brush. 1996. Anthropology and the conservation of biodiversity. *Annual Review of Anthropology* 25:329–352.

Pimbert, M. 1994. The need for another research paradigm. *Seedling* 11:20–32.

Plucknett, D. L., N. J. H. Smith, J. T. Williams, and N. M. Anishetty. 1987. *Gene Banks and the World's Food*. Princeton, NJ: Princeton University Press.

Rappaport, J. 1998. *The Politics of Memory: Native Historical Interpretation in the Colombian Andes*. Durham: Duke University Press.

Reaka-Kudla, M. L., D. E. Wilson, and E. O. Wilson, eds. 1997. *Biodiversity II*. Washington, DC: Joseph Henry Press.

Richards, P. 1986. *Coping with Hunger: Hazard and Experiment in a West African Farming System*. London: Allen and Unwin.

———. 1996. Culture and community values in the selection and maintenance of African rice. In *Valuing Local Knowledge: Indigenous People and Intellectual Property Rights*, edited by S. B. Brush and D. Stabinsky, 209–229. Washington, DC: Island Press.

Rigney, A. 2005. Plentitude, scarcity and circulation of cultural memory. *Journal of European Studies* 35(1):11–28.

Scott, James C. 1985. *Weapons of the Weak: Everyday Forms of Peasant Resistance*. New Haven, CT: Yale University Press.

Seremetakis, N. C. 1994. *The Senses Still: Perception and Memory as Material Culture in Modernity*. Chicago: University of Chicago Press.

Shiva, V. 1993. *Monocultures of the Mind: Perspectives on Biodiversity and Biotechnology*. London and Penang: Zed Books.

Soleri, D., and S. E. Smith. 1999. Conserving folk crop varieties: Different agricultures, different goals. In *Ethnoecology: Situated Knowledge/Located Lives*, edited by V. D. Nazarea, 8:133–154. Tucson: University of Arizona Press.

Soulé, M. 1993. Conservation: Tactics for a constant crisis. In *Perspectives on Biodiversity: Case Studies of Genetic Resource Conservation and Development*, edited by C. S. Potter, J. I. Cohen, D. Janczewski, 3–17. Washington, DC: American Association for the Advancement of Science.

Steinberg, M., and M. J. Taylor. 2003. Public memory and political power in Guatemala's postconflict landscape. *Geographical Review* 93(4):449–468.

Stephenson, David J., Jr. 1999. A practical primer on intellectual property rights in a contemporary ethnoecological context. In *Ethnoecology: Situated Knowledge/Located Lives*, edited by Virginia D. Nazarea, 230–248. Tucson: University of Arizona Press.

Stoller, P. 1994. Embodying colonial memories. *American Anthropologist* 96(3):634–648.

———. 1995. *Embodying Colonial Memories: Spirit Possession, Power and the Hauka in West Africa*. New York: Routledge.

Sutton, D. E. 1998. *Memories Cast in Stone: The Relevance of the Past in Everyday Life*. New York: Berg.

———. 2001. *Remembrance of Repasts: An Anthropology of Food and Memory*. New York: Berg.

Takacs, D. 1996. *The Idea of Biodiversity: Philosophies of Paradise.* Baltimore, MD: Johns Hopkins University Press.

Tonkin, E. 1992. *Narrating Our Pasts: The Social Construction of Oral History.* New York: Cambridge University Press.

United Nations Environmental Programme. 1994. Convention on Biological Diversity: Text and Annexes. UNEP/CBD/94/1. Geneva: Interim Secretariat for the American Biological Diversity.

Valladolid, J., and F. Apfel-Marglin. 2001. Andean cosmovision and the nurturing of biodiversity. In *Interbeing of Cosmology and Community*, edited by J. Grim, 639–670. Cambridge, MA: Harvard University Press.

Wilkes, G. 1991. In situ conservation of agricultural systems. In *Biodiversity, Culture, Conservation and Ecodevelopment*, edited by M. Oldfield and J. Alcorn, 86–101. Boulder, CO: Westview Press.

Wilson, E. O. 1988. The current state of biological diversity. In *Biodiveristy*, edited by E.O. Wilson, 3–18. Washington, DC: National Academy Press.

———. 1997. Introduction. In *Biodiversity II*, edited by M. L. Reaka-Kudla, D. E. Wilson, and E. O. Wilson, 1–3. Washington, DC: Joseph Henry Press.

Wolf, E. 1982. *Europe and the People without History.* Berkeley: University of California Press.

Zerubavel, Y. 1995. *Recovered Roots: Collective Memory and the Making of Israeli National Tradition.* Chicago: University of Chicago Press.

Zimmerer, K. 1996. *Changing Fortunes: Biodiversity and Peasant Livelihood in the Peruvian Andes.* Berkeley: University of California Press.

CHAPTER TWO

Apples of Their Eyes

Memory Keepers of the American South

SUSANNAH CHAPMAN AND TOM BROWN

A number of studies have revealed the ways that history, meaning, and memory are often embedded in the landscape and in particular places (Schama 1995; Feld and Basso 1996; Cruikshank 2005). A significant body of research demonstrates that long-lived trees often acquire cultural meaning beyond their immediate material uses (Dove 1998), while cultural memory can be embedded in the way that trees are named (Peluso 1996) or in the stories that accompany the transmission of different varietals (Nazarea 2006). A history of "the mundane" in the United States reveals the importance of apples and apple production in the daily lives of rural people. The thousands of apple varieties that flourished in early American history were well suited to the needs of a subsistence life: apples' versatility for fresh eating; drying; cooking; storage; animal feed; and producing cider, brandy, and vinegar made them an invaluable household staple.

During the nineteenth and twentieth centuries, radical transformations in household and agricultural production led to major changes in the ubiquity of apple diversity in the backyards of many American homes, with the concentration of much apple production into commercial orchards. While it is often assumed that concentration of production led to the erosion of crop diversity, the history of apple diversity follows a more complex pattern. Recent research on the changes in apple diversity in the United States during the twentieth century suggests that there is greater apple diversity commercially available today than there was just over one hundred years ago.[1] Such research challenges more common tropes about the loss of agricultural biodiversity and suggests a need to better examine the various sources of extant apple diversity in the United States today (Heald and Chapman 2012).

If the changes in apple diversity in the United States are not confined to an overall loss of apple diversity, questions then arise as to what changes in apple diversity have occurred, why such changes are so troubling to those who care about apple and other crop diversity, and what types of action are most useful for countering such changes. In this chapter we attempt to address these questions. The information presented here is based both on ethnographic research (S.C.) and long-term experience of searching for and collecting rare apple varieties in the southeastern United States (T.B.). Ethnographic research consisted of participant observation, life history, and semistructured interviews with a small network of nine apple collectors living in Alabama, Georgia, North Carolina, and Virginia.

The major change in apple diversity during the twentieth century has been a loss of the *ubiquity* of diversity. The alienation of apple diversity from the daily lives of most Americans has been accompanied by a parallel loss (and at times distortion) of memories and histories associated with such diversity. Here we define ubiquitous crop diversity as diversity that is salient and central to people's livelihoods and daily praxis. Rather than being concentrated in *ex situ* genebanks or *in situ* conservation orchards or gardens, it is mottled, patchy, and attenuated across a wide geographic area. It is diversity that is rooted in places, and thus it varies across the landscape—it is the family tree planted at the corner of a fence line or the upstart seedling planted by a bear in a clearing in the woods. Whereas old homesteads perhaps boasted a handful of apple varieties, those varieties varied from homesite to homesite, from town to town, and from region to region. These varieties were known and often cherished by children and elders alike, and the staggered ripening of different varieties marked the agricultural and social cycles of the year. Ubiquitous diversity is diversity that people use, tinker with, and transform, not only in its phenotypic characteristics but also in its practical uses and historical significance. It is characterized by what Eugene Hunn (1999) called *in vivo* conservation, or conservation through daily use. As such, ubiquitous diversity is more commonly situated within what Pierre Nora (1996) described as the *milieu* of memory than at any number of *sites* of memory (see chapter 1, this volume). Such "pied" and practiced diversity stands in contrast to the handful of apple varieties available today at nationwide grocery store chains or the concentrated diversity of historic conservation sites.[2]

Over the past thirty years, a small network of apple enthusiasts and small-scale nursery owners have begun to actively search for, propagate, and exchange old varieties and their associated histories. A number of the apple varieties found by apple collectors have made their way to historic sites, while others have wound up in personal or community orchards. People who remember and cultivate these apples are involved in memorial practices that are at once deeply personal, social, and cultural. The work of "apple collectors," the memories of the people who live amongst many of the old apple trees dotting rural landscapes, the importance of apples to local and cultural identity, and the longevity of old apple trees have been the primary forces countering the potential loss of old-timey apple diversity. The work of apple collectors has been, and still is, central to processes aimed at reestablishing the ubiquity of apple diversity. Today old apple trees dot rural landscapes as markers of homesteads, orchards, and communities—markers that remind current residents of times past. But the apple trees in the landscape and the fruits that they provide also serve as repositories of personal and cultural memory. It is through the collection, propagation, and exchange of these various apple varieties that the work of apple collectors has enabled the (re)construction of new social meanings associated with old varieties and the production of new memories.

The work of apple collectors has thus been an important component of blurring the boundaries between the milieu of memory and sites of memory and of supporting both *in vivo* conservation and *in situ* conservation of apples, a muddling that is necessary not only for the long-term conservation of older apple diversity but also for the future creation of new apple diversity. In particular, the work of apple collectors is as much concerned with the conservation of extant diversity as it is with the maintenance and creation of the processes that gave rise to such diversity. As such, the importance of their action comes less from the absolute numbers of varieties conserved or created by their work than it does from the type of diversity, the everyday histories, and the lived memories that their work produces and maintains.

The first section of this chapter introduces the history of apple diversity in the southeastern United States, tracing the shift from ubiquitous apple diversity to what could be characterized as the striking absence of apple diversity in the daily lives of most Americans. The second section explores the relationship between apple diversity and memory and focuses on the ways in which the nomenclature of apples serves as a repository of cultural, historical, agricultural, and culinary knowledge. As such, naming has been an important component in the relocation of apple varieties, in the maintenance of knowledge of different cultivars, and, at times, in the transformation of social meaning attributed to different varieties. The third section explores how apple varieties can become movable "places," which, through their exchange and propagation, can serve as bearers of personal, social, and cultural memories. Apples serve both as cultural markers on the landscape and as ways in which humans actively imbue places with identity and history through the processes of everyday conservation. The final section traces how the work of apple collectors fosters ubiquitous apple diversity via the maintenance of processes that aid the (re)construction of "place": exchange of seedling and grafted apple varieties and the circulation of practical, personal, and cultural memories of apple cultivars. These processes are often evident in the ways that different varietals are categorized and acquire new social meanings, such as heirloom or heritage status.

Apple Histories

The history of domesticated apples (*Malus domestica*) in the United States tells a story of proliferation of varietal diversity—diversity created and maintained through the planting of seedling orchards, the exchange of varietals via small nurseries and apple peddlers, and the requirements of daily use. *Malus domestica* was originally domesticated in the Tian Shan range that straddles present-day southeastern Kazakhstan and western China (Harris et al. 2002; Juniper 2007). While the Americas possessed many native fruits, the only *Malus* species present when European colonists arrived were hard, almost inedible crabapples. By the early 1600s, specimens of *Malus domestica* had been imported to the American colonies by way of England and France. As a result, much of the earliest imports of apple germplasm to the colonies and early United States were varieties loved by the earliest European immigrants (Lape 1979).

Prior to the American Revolution, propagation of fruit trees by grafting was not widespread, as the grafting of fruit trees—the clonal propagation of a prized variety by inserting a cutting into the root system of another tree—was not a widespread skill (Diamond 2010). European settlers without access to grafted stock were planting apple seeds and root sprouts (Beach 1905; Hedrick 1950).[3] Some early colonial policies even mandated that individuals who were granted sizable parcels of land plant orchards of fruit (Hensley 2005), while in other cases tenant farmers were required to plant orchards as part of their lease agreement. Most often these orchards were seedling orchards, as seeds were more affordable for the small farmer than grafted stock or root sprouts. Because apples possess a high degree of heterozygosity in their genetic recombination, the fruit of young apple trees grown from seeds rarely resemble the fruit of their parent trees. The planting of seedling orchards thus fostered the ecological and economic conditions amenable to the wide-spread selection of new apple varieties. Creighton Lee Calhoun Jr. has described the profusion of seedling orchards from 1607 to 1900 as "literally a vast agricultural experiment station, crudely screening apple varieties, rejecting most but keeping the best" (1995:9).

In the years following the American Revolution, changing foodways and dietary patterns allowed for the increased consumption of fruits and baked goods. People began to seek out renowned apple varieties, prized for their flavor, texture, and suitability for various everyday uses. The early nineteenth century also witnessed the rise of fruit tree production as a specialized agricultural endeavor, and by the mid-1800s a number of tree nurseries were scattered throughout the eastern United States specializing in the propagation, selection, and maintenance of desirable apple varieties (Diamond 2010). When people identified excellent young seedling apples, the trees were grafted, named, and shared with family and neighbors or added to nursery catalog listings. Exchange of named apple trees through peddlers, small nurseries, and community ties ensured that useful varieties, and often their stories, were shared among households, towns, and regions. Unless seedling apples are clonally propagated, a unique variety will vanish when the original tree dies. The exchange of root sprouts and grafted apple trees thus ensured a widespread distribution of cherished apple varieties—where names, stories, and memories were passed along with the tree.[4] This combination of the habitual planting of seedling orchards and clonal propagation for varietal exchange enabled the emergence of incredible apple diversity in the early United States and was accompanied by the spread of many named varieties to areas outside of their place of origin.

During the twentieth century, rural outmigration coupled with changes in the organization of agricultural production led to major transformations in most Americans' relationship with apple diversity. Fewer people were relying on household and local production. Fowler and Mooney (1990) noted that large-scale commercial production for emerging grocery store chains became concentrated around a handful of crop varieties. This shift is reflected in the types of varieties cultivated for wholesale markets. More recently, Goland and Bauer (2004) found that orchardists in Ohio distinguish between apples grown for wholesale commercial markets and those grown for local "niche" markets. An apple variety's appearance, resistance to bruising, and keeping quality are all factors that limit the varieties growers choose to produce for larger commercial markets. Many of the older varieties prized for flavor or home use do not meet the requirements for large-scale production and shipping. The Red Delicious apple was discovered in the late 1800s, and during the twentieth century it became one of the most prized apples for commercial fruit production. As one apple collector from Georgia noted, "All the fruit breeders keep trying to come up with something that gets red quick to try to beat the market. All these new strains of Red Delicious, they look ripe, but they're not. [They want] something tough enough to ship. But that Tenderskin [apple], now, there is no way you can ship it. It would get bruised up. . . . It's a mighty good apple. Better than any you can find on the market."

Small growers struggled to compete with large-scale orchards as wholesale apple production became more prevalent throughout the country. In 1921, Folger and Thomson noted a decline in farm orchards. They wrote, "at present commercial apples can be grown successfully only when scientific and intensive cultural methods are employed. The farmer can no longer give his orchard indifferent care and expect to compete with the commercial grower" (1921:3). Implicit in Folger and Thomson's account is a valuation about apple quality amidst changing production and consumer standards. Apple varieties that did not possess a long shelf life were less desirable for long-distance markets that required exceptional keeping qualities in fruits and vegetables. Less-than-perfect apples were losing out among consumers at the grocery store as well, as visual appearance—rather than taste, smell, or feel—was becoming a primary criterion for purchasing fruits and vegetables for the average store clientele.

Many apple collectors, who continue to grow, propagate, and sell old-timey apple varieties, also note this shift. Tim Hensley runs a small and diverse fruit and nut tree nursery in Bristol, Virginia. During an interview about his personal experiences collecting and propagating different apple varieties, he discussed how the loss of subsistence farming and the rise of suburban landscape values affected apple diversity. According to Hensley, "If you look at the old garden magazines from the '60s and '50s, maybe the '60s especially . . . it's like almost a sterile landscape—that messlessness seems to be one of the ideals, and you can't have no mess with fruit trees." Changes in production and changes in the ideals about landscape, which were often intricately tied to changing class consciousness, contributed to the transformation of the ubiquity of apple diversity during the twentieth century. Production, once centered on diversity, subsistence, and local markets, was gradually replaced by large-scale commercial orchards, grocery stores, and a handful of apple varieties suited to the new production environment (see chapter 12, this volume).

In his 1905 publication, *Nomenclature of the Apple*, W. H. Ragan lists an impressive 7,000 different apple varieties known to be grown in the United States during the nineteenth century (for excellent accounts of this history, see Calhoun 1995; Hensley 2005).[5] Despite the high degree of diversity documented by Ragan for the nineteenth century, it is estimated that somewhere between 280 and 420 distinct apple varieties were offered for sale in 1905 commercial nursery catalogs (Heald and Chapman 2012). These numbers probably do not reflect a catastrophic loss of apple diversity at the start of the twentieth century. Instead, it is likely that in no single year during the nineteenth or the twentieth century were all 7,000 "Ragan" varieties offered for sale in nursery catalogs. Rather, different varieties of apples were added and dropped from nursery catalog listings in any given year (Heald and Chapman 2012).

By the end of the twentieth century, at least 1,469 apple tree varieties, many of which are "old-timey" varieties, were commercially available in American nursery catalogs (Heald and Chapman 2012). This figure is a stark contrast to the number of apple varieties currently grown for commercial wholesale fruit production in the United States, which focuses on the cultivation of approximately ninety varieties (Hensley 2005). Likewise, in 1995, Lee Calhoun, an apple collector and orchardist, estimated that of the 1,400 varieties of apples known to have originated or have been widely grown in the southeastern United States, fewer than 800 varieties were known to still exist. Since 1995, he and other collectors of old-timey apples have relocated hundreds of these "extinct" varieties, as well as a number of varieties that never made the ranks of historic variety registers, apple lists from the U.S. Department of Agriculture (USDA), or published nursery catalogues. The relocation of these varieties has succeeded because old apple trees, the people who remember them, and the people who care to seek them out still exist. Together with USDA variety registers and old nursery catalog listings for some varieties, these trees and people provide the invaluable mixture of historic documentation, personal and cultural memory, and rare germplasm that have made the maintenance of crop diversity possible.

Although changes in rural livelihoods and agricultural production over the past century have posed real dilemmas for the maintenance of crop diversity, research in anthropology shows that diversity has persisted in farmers' fields, kitchen cupboards, homegardens, local markets, and even the collections of small commercial seed and nursery companies (Nabhan 1989; Nazarea 1998, 2006; Brush 2004; Veteto 2008). Despite this persistence of crop diversity "in the margins" (Nazarea 2006), the loss of ubiquitous diversity is troubling for a number of reasons, many of which are well documented in the existing literature on the erosion of crop diversity. These include the loss of cultural knowledge, memories, and histories associated with crop diversity (Nazarea 1998, 2005); the genetic selection acting on *ex situ* germplasm that favors the conditions of *ex situ* preservation over the conditions of farmers' fields (Soleri and Smith 1999); and the extraction of crop diversity from the *in vivo* conditions that foster those social and ecological processes that create crop diversity (Hunn 1999). Perhaps the most troubling aspect of the loss of ubiquitous diversity is the possibility of absolute loss as old apple trees and the people who tend them, use them, and remember them pass away. However, recognizing the decline of ubiquitous diversity—in this case with apples—questions narratives of absolute loss and forces scholars and farmers to consider new possibilities for conservation and the creation of new diversity.

Apple Names and Cultural Memory

Numerous scholars have demonstrated the important relationship between cultural memory and crop diversity (Richards 1996; Nazarea 1998; Dove 1999). This body of research highlights the ways that cultural memory and biodiversity guard against the loss of options, customs, and traditions amidst changing social and ecological relations of agricultural production. Such work emphasizes the dynamic processes that contribute to the maintenance of crop varieties and their associated memories (Nazarea 2006). Just as the colors, the smells, and the flavors of fruits all aid in the identification of old trees, so too have the memories of different apple varieties helped corroborate the identity of old trees growing in orchards and homesteads. This relationship among apple naming patterns, cultural memory, and personal recollections has been an important factor in the maintenance of extant diversity and the relocation of lost diversity.

Research on varietal naming has documented the patterns farmers commonly use to name landrace and folk varieties. Varietal names can under- or overrepresent varietal diversity (Sadiki et al. 2007). Varietal names reflect the ways humans classify varieties based on broader social, cultural, and economic values. Varietal names can be basic or nonbasic (Hmimsa et al. 2012). Basic names are based on the morphological and agronomic characteristics of a variety, or they may reflect the social history and geographic origin of a variety. By comparison, nonbasic term names are seemingly arbitrary or whimsical, meaning they do not impart agronomic, historic, social, or geographic information about the variety. Basic-term names provide information to apple collectors that aid searches for "lost" varieties—particularly when little written or oral information about the apple exists. Such apple names commonly provide information about the visual appearance of a fruit: fruit color, skin texture and thickness, and fruit shape. In Wilkes County, North Carolina, examples of apple names include the Speckled Red, Red Streak, and Greenskin apples. The Yellow Meat apple of the same county has a pronounced yellow flesh, while the White Horse has a yellow skin but very white flesh. Other apple names that reflect external characteristics include Tenderskin, Thinskin, Toughhide, Leather Coat, Greasy Coat, and Knobbed Russet, as well as the very common moniker Rusty Coat, which is given for the russet on an apple's skin. Names such as Choking Sweet, Water Core, Juicy, Greasy, and Tobacco Sweet allude to texture, color, or even sensitivity to oxidation, while names such as Sheepnose, Crow Egg, Black Banana, Pear, Barrel, and Biscuit reflect the shape of the fruit.

Still other names such as Sweet Dixon, Sour Russet, and Bitter Buckingham impart information about the flavor of fruits and their preferred uses. The name Sour Sweetening reflects the way the apple's flavor changes during the ripening process. Before fully ripening, the Sour Sweetening apple is so sour and bitter that it is inedible, but when it is fully ripe it becomes quite sweet and good for fresh eating. Parts of names such as "Cider," "Jack," and "Keeper" hint at the popular uses of different varieties, although in the past few apple varieties were used for only one purpose. Still, some apples were better suited than others for producing certain foods. For example, some apple varieties, no matter how long one cooks them, will never dissolve into applesauce or apple butter, while other apple varieties will become too soft when cooked into pies. In other cases, memories about different varieties provide important culinary or processing information, even if such qualities are not embedded in the apple name. In Wilkes County, for example, the Payne apple was used to make the best-tasting apple brandy, but the Yellow Hardin would produce the most brandy per bushel of apples. Table 2.1 gives an overview of some of the cultural and horticultural information that can be embedded in apple names.

Apples also can be named after people who developed the new variety, owned the land on which the new variety originated, introduced the apple into a new area, or are associated with the apple through an interesting event. Stories link a number of old apples to different people and places. Apple names such as Sally Crockett, Beck Searcy, Polly Cook, and Junaluskee all reference the people who played a role in the origin, discovery, or development of the variety. Ron and Suzanne Joyner, who collect and propagate old-timey apples and run an apple nursery in Lansing, North Carolina, obtained access to a seedling tree near their nursery, grafted it, and named it after the woman on whose land the tree was located and who had used the apples for years. During one interview with the Joyners, Ron recounted:

> The Clara's Creek apple was one we found on the other side of the county with this really sweet old lady. We met her, Clara Dougherty. She was ninety-four years old when her son contacted us and said his

Table 2.1. The types of information embedded in apple names

Category	Examples of Apple Names
Skin	
Color	Greenskin, Arkansas Black
Texture	Knobbed Russet, Roxbury Russet, Rusty Coat
Thickness	Thinskin, Toughhide, Leather Coat
Sheen	Greasy Coat, Greasy
Pattern	Red Streak, Speckled Red
Flesh	
Color	Snow Apple, Yellow Meat, White Horse
Texture	Choking Sweet, Juicy Queen
Oxidation	Tobacco Sweet
Flavor	Sweet Dixon, Sour Russet, Bitter Buckingham, Sour Sweetening, Husk Spice
Use	Smith's Cider, North Carolina Keeper, Melt in the Mouth
Other	
Shape	Sheepnose, Summer Ladyfinger, Crow Egg, Black Banana, Barrel
Size	Little Benny, Red Pound, Mammoth Pippin, Ten Ounce, Twenty Ounce
Ripening time	Early June, Early Ripe, May apple, Oat apple, Fall apple, McCuller's Winter, Forward Sour
Tree shape	Limbertwig, High Top, Bushy Top, Tall Top
Bloom patterns	Blackberry apple, Double Blooming
People	Junaluskee, Sally Crockett, Beck Searcy, John Connor, Billy Spark's Sweetening
Places	
Near-by objects	Hollow Log, Bee Bench, Woodpile, Hog Pen, Foot Log
Tree location	Poorhouse, Polly Sweet (located at the home of Polly Cook)
Events	Appomattox Golden Sweet

mother had this old apple that was getting old(er) and they wanted to save it and plant a few more trees. . . . It was on the creek and she loved the apples, but they would fall in the creek and she would lose them . . . so we named [it] Clara's Creek apple. . . . It is this small yellow apple, not very exciting to look at, but [it has a] wonderful flavor— this really warm rich sweetness to it. [Clara] liked it for cooking and frying—what the other people up here call "fruit n' eggs." It was fried apples and eggs.

Apples can be named after their place of origination or an unusual event related to their discovery or exchange. One apple collector in Georgia recalled grafting the Duck apple for a local man. He said the man had brought cuttings of the tree, which had grown at his grand-mother's home. "The tree was really tall. An apple fell out, hit the duck in the head, and killed it. They named it the Duck apple." In other cases, such names inspire the maintenance or relocation of old apples by tying the variety—and those who keep it—to particular social, political, and historical events. According to one story, the apple called Appomattox Golden Sweet was found by a Civil War soldier near Appomattox, Vir-ginia, as he returned from war after General Lee's surrender. The sol-dier's grandson still tends an Appomattox Golden Sweet apple tree.

A number of old apple varieties have been named or memorialized for events and people in national history. Such varieties include the Esopus Spitzenburg, heralded as Thomas Jefferson's favorite apple, and the Rox-bury Russet apple, which has a history dating back to just twenty years after the arrival of English settlers at Plymouth Rock. The story of an-other famous old apple, the Ralls Janet, links the variety to early Ameri-can politics. According to one version of the story (of which there are many), the French ambassador to the United States during George Washington's presidency, Monsieur Genet, introduced cuttings of the tree into the United States. Thomas Jefferson, secretary of state at the time, admired the apple so much that he acquired cuttings from Mon-sieur Genet and encouraged the variety's propagation throughout Vir-ginia. Jefferson then gave cuttings of the tree to Mr. Rawles, a local nurseryman and acquaintance. The tree was later renamed Ralls Janet, in commemoration of Mr. Rawles and Monsieur Genet. Calhoun (1995) notes that there is no evidence that the Ralls Janet was brought to the United States by Monsieur Genet. By linking the history of the Ralls Janet to historic events, fabricated or not, the story authenticates the age and cultural importance of the variety.

While apple synonyms can complicate the quantification of apple di-versity and the identification of apple varieties, they also mark the layer-ing of social and cultural memory onto certain cultivars. Indeed, famous apples may have many synonyms, some of which are only known in the locality of a cherished tree of the famous variety. This circulation and layering of memory has significant implications for the way that apples are categorized and ascribed social meaning. While some apples have achieved the status of famous heirloom or heritage variety, other old varieties may be known only locally. All apple collectors distinguished between seedling, local, or family varieties and those apples that were regionally or nationally famous. Ron and Suzanne Joyner explained the difference between "known, documented varieties" and "apples that just came up as seedlings in someone's back pasture or their front yard." If the "yard" apple "turns out to be a good variety, [people] keep it. . . . These yard apples are shared with family [and] sometimes [people] will propagate it and share it with other members of their community."

The ways that memories layer upon some apple varieties and the ways that synonyms facilitate such layering have important implications for the social classification of apples. In some cases, the synonyms attributed to a single variety tie the legacy of a variety to very different histories and serve to reproduce various personal, social, or cultural memories. The cultural memories attributed to different yard apples or famous varieties can depend on a number of factors: the histories and the stories attached to the variety, the geographic range of the variety in the past and the present, written accounts of the variety in historic documents, and, in some cases, the quality and potential uses of the apple. All of these factors intersect to influence the social meaning attributed to an apple.

The memorialization of relatively famous apple varieties has played a crucial role in the maintenance of those varieties, as people know enough about them to know they are "lost," to search them out, to select those trees to plant in their gardens and orchards, and to incorporate them into heritage sites. But it is also the ability of apple varieties, their trees, and their fruits to instill a sense of place and to provide links to personal and localized communal acts of remembrance that helps ensure that the apple diversity passed on is tempered by more inclusive forces. By "inclusive forces," we mean those processes that inspire people to maintain, care for, and pass along the lesser known old varieties—"yard" and seedling apples—the apples that never appeared in historic documents, the old hollowed-out apple trees that mark yesterday's homesites, and the fruit of seedling apples fried in hog fat that people remember eating at their grandmothers' kitchen tables. It is these same "inclusive forces" that inspire people to cherish and pass along "new" varieties—those varieties that, due to their vintage, are not yet considered heirloom. As one collector noted, "I don't care if it is a known variety and if it is written in the history books. If it is a good apple, it is worth saving."

Apple Diversity and the Making and Remaking of Place

Apple trees, in their rootedness and in the ease with which they can be shared with others, are at once places in themselves and conduits through which people create ties to distant locales, departed relatives, and events of the past. The histories and memories associated with apple varieties and their past places often inspire others to maintain, collect, graft, and exchange rare varietals. The people involved in collecting these apples share a curiosity about old varieties that carry historical significance in a particular region, as well as a desire to propagate and retain old local or family apple trees. As such, all apple collectors annually grafted a large number of locally famous trees—for which they had many requests—as well as a number of lesser known varieties.

During an interview, Steve Kelly, a nursery owner and apple collector in Big Stone Gap, Virginia, discussed the apple varieties he has collected and grafted. He explained that many of the apples he collects are strains of varieties that are common around Big Stone Gap:

> Most of the ones that I collect around here, I probably already have the variety but [they're from] old-timers that I knew and I want it just because it belonged to them. And you know, I don't care if you get [one] from here or one from over there, lots of times they are different. When I first started out [I thought], "why I've already got that, I don't need it." So I didn't get it. But later on I found out that some have different characteristics. . . . I've got several different strains of Benham. And a lot of people [who have] grown them for years, they know the difference.

He explained that "somebody out there one day may want one [apple variety] that their Granddaddy had and then maybe I'll have it. It won't be lost if I can help it, the old varieties anyway."

Tim Hensley of Bristol, Virginia, first got started grafting old apple trees while he was working at a nearby commercial fruit nursery as a young man. People would come into the nursery asking for different old apple varieties that the nursery did not carry because such varieties were "not propagated in quantities by big growers." The public's curiosity inspired him to begin looking for old apple varieties. Soon thereafter, he began taking orders for his own nursery business. He recalled, "[I] had to scramble at first to fill the orders. And of course then I had to start learning because folks wanted to know 'what's a Buckingham?' And what is a Buckingham unless you get in there and dig out some of the history that you don't know?"

Memories associated with apples can be both immensely personal and cultural. Personal memories about old apple varieties link people with those who have gone before them, shared childhood and communal events, and places meaningful in their own pasts. Cultural memories, on the other hand, provide people with a sense of belonging and a feeling of broader membership in a real or imagined community. Ann Rigney describes cultural memory as "working memory" that is mutable, transitive, and "continuously performed by individuals and groups as they recollect the past selectively through various media and become involved in various forms of memorial activity" (2005:17). With apples, recollection occurs through sharing and exchanging fruits, stories, and trees, while the "media of transmission" for cultural memory include written descriptions, oral accounts, and, of course, the apples themselves. The sensory experience of tasting, smelling, or touching an apple—or the act of savoring a slice of apple pie at grandmother's kitchen table—suggests that memories surrounding apples can be both social and immensely personal (Lupton 1994). Such is the ability of food to create meaning that transcends the immediate needs of everyday life (Holtzman 2006). Steve Kelly discussed one tree he got from a relative. He said, "[I found it at] my grandmother's place, my great-grandmother's place down the road here. Bailey's June: I believe I've got a picture of it, too. She used to make fried pies out of it when I was a kid, and I hope it's the same one. I'm not sure yet, but we got cuttings from it and I had to wade through the briars to get it."

Personal and cultural remembrance practices associated with apples are also a means to construct *new* places and imbue them with meaning. The ability of apples to serve as a link to social relationships, historical events, and places of the past is a key factor in the production of "senses of place." Edward Casey defines *place* as "named or nameable parts of the landscape of a region, its condensed and lived physiognomy" (1996:28). Indeed, place names are often concentrated meaning—reflecting the symbolism and cultural memory that has been imbued in a site (Basso 1984; Carter et al. 1993). In many cases, the layers of memories and meanings instilled in apples, along with the rootedness of the tree itself, make apple trees places in their own right, and the ability of a particular variety to spark memories of the past fosters the construction of senses of place.

All apple collectors said that they regularly have people approach them to graft an old family apple tree. Ron Joyner recalled, "Every winter we get dozens of people sending us cuttings in the mail, sometimes little cuttings, [but] we'll also get four foot boxes with whole tree limbs and they'll say, 'give me as many as you can get of this.'" Similarly, Joyce Neighbors, who runs a nursery in Gadsden, Alabama, and who has collected old-timey apple varieties for the past thirty years, noted that many of the people who buy apple trees from her are searching for varieties that they remember from when they were younger, varieties that their families kept and that create a tangible link to past relationships. She noted that people want a variety that they recognize from their childhood, recalling that "people will say 'I want it because it was on my granddaddy's old place and I always loved to go see granddaddy and get some of those old apples.'" In her experience, "it was the memories [of the old apples] that were drawing people" to plant old apple varieties.

While people might invest cultural and personal memories in a particular variety of apple, a tree can also be imbued with personal and experiential memories specific to "that tree." Clonal propagation of apples muddles this relationship a bit, since apples instilled with the memory of one place can be moved to another. In the case of heirloom seeds, Virginia Nazarea reminds us that "the long and intimate history of these seeds makes them perfect vehicles for transporting the sights, smells, and textures as well as emotions, stories, and memories that make for a vibrant sense of place" (2005:80). Thus, in their mobility no less than in their rootedness, apples serve as repositories and vehicles for memories and senses of place.

Apple diversity can in itself be symbolic of particular places. In Wilkes County, people have reiterated the importance of keeping different varieties of apple trees than their neighbors. Having distinct and unusual apple varieties was a point of pride that distinguished one homestead from the next. Today, not only do most old homesteads have a diversity of apples, but the apple varieties that have been found growing in Wilkes County tend to be rare old varieties, since from one homestead to the next, people kept trees of different varieties. This contrasts sharply with the neighboring county, Alleghany, where many households planted the same varieties as their neighbors, and a majority of the apples extant there today are common old-timey varieties. As landmarks of former places, remnants of old home sites, and markers of once-prolific orchards, apple trees serve as markers of the land's history.

Simon Schama argues "before it can ever be a repose for the senses, landscape is the work of the mind. Its scenery is built up as much from strata of memory as from layers of rock" (1995:6–7). Apples engender a sense of place by aiding the remembrance of the past, as living reminders of past relationships or membership within a broader society. They also serve as markers of passed-on places, of the imprint on the land left by people who have now passed away. The importance of apples to the production of place has not only helped to maintain high levels of diversity—as in Wilkes County, North Carolina—but also it has encouraged individuals to search for, collect, and propagate old apple varieties that reconnect them to their own past or to a past that is mediated through shared cultural memory.

Circulation and the "Milieu"

The planting of apple trees, the consuming of different apple varieties, and the recollection of old and creation of new memories are all important personal and social practices that help resituate diverse, rare, often relatively unknown apple cultivars into the milieu of memory. Likewise, the memorial practices associated with an agrarian American past have been instrumental in the creation of conservation and heritage orchards. Mutually reinforcing, these types of activities have inspired the maintenance, propagation, and (re)location of many old-timey apple varieties. Such personal and cultural memorial practices are different parts of a reiterative process—they are not separate actions, and the work of apple collectors is testament to this muddling. All of the collectors and growers who participated in this study have been involved in (re)locating the well-known heirloom varieties that were at one time thought to be extinct, such as Reasor Green, Carter's Blue, Bryson's Seedling, North Carolina Beauty, Bunker Hill, or the Junaluskee, to name but a few. At the same time, all of these apple enthusiasts have also been involved in propagating cherished, though less famous, "yard apples." Apple collectors have even been involved in propagating and exchanging unknown seedling apple trees, such as those "planted" by bears in the mountains. Ron and Suzanne Joyner, for instance, found, grafted, and named two varieties—the Husk Spice and the Husk Sweet—in just this manner.

During interviews, all apple collectors emphasized the importance of exchange and circulation for the maintenance of apple diversity. As collectors locate "lost" varieties, yard apples, or unknown seedlings, they propagate the variety, and eventually share cuttings with other collectors and the general public. Ron Joyner explained that "everybody in this network realizes it is more important to propagate these, keep these varieties living, keep them going, rather than hoarding them. We are the only ones who have the [Bunker Hill apple], but eventually everybody in this network will have it." Steve Kelly recounted the story of locating a healthy Raleigh apple tree—a rare old variety that was in the collections of some orchardists but was diseased. He said, "I got the wood and I sent it all over the United States. I sent it to California, Oregon, Kentucky, to all the collectors that I knew. I told them to pass it on to the ones they knew. So that's one that we got back in circulation." In addition to sharing cuttings with other apple enthusiasts, collectors will often send cuttings to heritage sites and preservation orchards. This exchange and circulation of rare apple varieties is reminiscent of the historic exchange of cultivars among peddlers, small nursery owners, families, and individuals that fostered ubiquitous diversity.

Through varietal exchange and movement of trees, new memories are attached to apple cultivars that influence the social attribute—heirloom and heritage, yard, family, and seedlings—each receives. Yet the social meaning attributed to different apple varieties is never permanent. The social meaning of an apple variety can be transformed in a number of ways, such as through propagation and exchange, renaming, and the creation of new memories that are attributed to the variety as it is shared and experienced by individuals and groups. It is through this transformation of the social meaning of apples that the blurring of the milieu of memory and sites of memory is most evident and when the work of apple collectors is exceptionally significant. This is best illustrated by the story of the Junaluskee apple (sometimes called the Junaluska or Junaliska apple)— that just fifteen years ago was thought to be extinct.

The story of the Junaluskee apple was recounted by Silas McDowell, a nineteenth-century nurseryman who first began grafting the tree for sale. The original seedling tree was located on the land of Cherokee leader Junaluska, who resided in what is now either Macon or Cherokee County, North Carolina. When the state commissioners arrived to purchase this portion of Cherokee territory from its residents, Junaluska refused to part with the land where his favorite apple tree was growing. Eventually, the state commissioners agreed to pay Junaluska an extra $50 for his apple tree. After the state purchased the land and the tree, it is likely that the tree was not propagated until McDowell, after learning of the story of the seedling tree, tracked down its location. He subsequently took cuttings from the original tree, propagated it, and through his personal fruit tree nursery began to popularize the story of Junaluska's favorite apple (Calhoun 1995).

But with the 1995 publication of Lee Calhoun's *Old Southern Apples*, the Junaluskee apple was listed as one of the old apple varieties thought to be extinct. After years searching for a living Junaluskee tree, apple collector and coauthor Tom Brown met a man in Franklin, North Carolina, who knew of an apple tree that fit the description of the Junaluskee apple as it was described in Silas McDowell's nursery catalog. The man accompanied Tom to the tree, which was known locally as the "John Berry Keeper." Macon County residents referred to the tree as the John Berry Keeper because it grew at John Berry's old home site and the fruit from the tree kept well through the winter. The John Berry Keeper was eventually identified as the Junaluskee apple by cross-referencing the apple with early USDA descriptions and by obtaining positive identifications from four other people who remembered the Junaluskee apple from their childhood. After obtaining a positive identification, Tom Brown propagated the tree. He subsequently donated three Junaluskee apple trees to the Junaluska Memorial and Museum in North Carolina, where they were planted next to Junaluska's grave site. He also shared cuttings of the tree with other apple collectors for further propagation—a common practice within this network of apple keepers. Today, a number of different historic sites and heritage apple projects boast specimens of the Junaluskee apple in their orchard listings, as do, undoubtedly, a number of home orchards.

It is uncertain whether the history of the original Junaluskee tree is accurate or apocryphal, or a little of both. The story does appeal to certain cultural tropes associated with recollections of early American history. Particularly interesting is the type of agency and resistance shown by Junaluska to the state showing up at his home to "purchase" his land. His ability to successfully demand an additional fifty dollars from the U.S. government—no small sum in those days—is testament to his attachment to his home and to his apple tree, but it is also a victorious account of Indian resistance amidst many more horrific events in the history of U.S.–Cherokee relations. The fact that the apple tree was Junaluska's successful bargaining point only amplifies the cultural and historical significance of the tree and the apple's ability to create a living link to such a victory. It also suggests that, perhaps, the Junaluskee is *some apple*, not only one that Junaluska prized, but also one whose worth was subsequently recognized by the government commissioners.

Thus, as the Junaluskee apple began as a cherished seedling on Cherokee land, an interesting historical exchange helped set the variety apart in the social imagination of the region's residents. That difference inspired McDowell to search for and find the original tree, after which point he popularized the tree and its history through his own nursery business. Because the variety was old and was documented in historical publications—in McDowell's own nursery catalogs and subsequent USDA descriptions—it eventually came to be considered a famous heirloom apple. Ultimately, the Junaluskee was so well known that people realized it became "lost." But, of course, the tree was not ever entirely lost—it had only been exchanged, renamed, and given new localized meaning, as the John Berry Keeper.

The journey of the Junaluskee apple from yard apple to heirloom apple (and back again as the John Berry Keeper) demonstrates that the processes associated with different memorial activities are in no way separate. Both personal and cultural memorial processes contribute to the construction of place, although in different ways. Both types of memorial processes can be situated within the milieu of memory and can be experienced at sites of memory. Indeed, quite personal recollections do occur at sites of memory; a person, for example, who knows the Junaluskee apple as the John Berry Keeper may have a very different recollective experience at the Junaluska Memorial and Museum than does another person who visits the site and meets the apple for the first time as the Junaluskee.

Perhaps the greater implication for apple diversity is that the milieu of memory, the memory that "sits in places" (Basso 1984) and is (re)created in the everyday, provides much more opportunity to mix things up, to remember, to forget, to make new memories, to create and remember ties among apples, people, events, and places, to forget it all again, to find and exchange new apples and new memories, and of course, to lose it all again. There is certainly a part of this broader milieu of memory and *in vivo* conservation that is quite vulnerable to the types of social and agricultural changes that occurred during the twentieth century. But, there exists also the recombinative capacity of the milieu, in its diversity of everyday places and memories, that makes it ultimately more creative.

Apple collectors have played an invaluable role in this creative process. Not only are they searching for famous old varieties, memorialized in old publications and oral histories, but in their searching they are also collecting a number of amazing varieties and strains whose greatest significance is to the people on whose land the tree stands. At other times, they may be propagating, naming, and circulating apple varieties that no one previously knew, thus adding new or relatively unknown varieties to the roster of existing apple diversity. In their push to get all of these varieties into greater circulation, collectors are helping build the diversity of heritage orchard sites and are providing the homegardener with an incredibly diverse selection of old-timey (and sometimes new) apple varieties. It is in this sense that they are not merely conserving diversity but helping to perpetuate the memorial and creative processes that enabled the rise of immense and ubiquitous apple diversity in the first place.

(Re)Placing Ubiquity

Although the twentieth century witnessed the loss of ubiquitous apple diversity, it did not witness an entirely catastrophic loss of gross apple diversity (Heald and Chapman 2012). The power of apples to provide a sense of continuity to past events, people, and places has inspired people to seek out old, "lost," and excellent varieties. This is not to say that much of the rich knowledge and memories associated with apples has been successfully transmitted to younger generations. Indeed, one trend reiterated by apple collectors was the problem of the passing of the elderly who had knowledge of and experience with apple diversity and who knew the appearances, the tastes, the uses, the people, and the events surrounding old apple trees and apple varieties. Rather, much of the maintenance and recovery of apple diversity has occurred where people have decided to actively search for famous old apples thought to be extinct, to graft dying trees from their old friend's home site, or to find and plant an old variety that their grandparents kept. What is involved in all of these actions is an engagement with the memories of the past—whether personal or cultural—and the actions that reproduce memory in the everyday.

Apple collectors are memory keepers and memory makers—as open to finding famous old varieties as they are to multiplying relatively "unknown" varieties or strains. Many are not afraid to recognize the occasional value of a "spitter," and as one apple collector put it, "Who knows, maybe a horse will eat it!" Apple collectors' impact on apple diversity has been unique and remarkable. As a group, they have not only relocated dozens of the varieties that Ragan documented but also found hundreds of old and perhaps not-so-old varieties that never were recorded in historic nursery catalogs, USDA lists, or other publications.[6] Over the last ten years, coauthor Tom Brown has found and collected the fruit of 920 different apple varieties. As apple collectors locate new apple varieties, they propagate trees, collecting and disseminating valuable memories, stories, and histories of those varieties in the process. Such continued action is necessary not only for the maintenance of extant apple diversity but also, just as important, for the creation of new apple diversity.

Notes

1. This measure of commercially available diversity is based on the diversity of apple tree varieties commercially accessible to the American public. The study does not attempt to measure the diversity of commercially available apple fruit varieties, which is presumably quite different, if not abysmal, based on the selection at most major grocery stores. For more information, see Heald and Chapman (2012).

2. A number of preservation orchards for apple diversity have emerged in the last twenty to thirty years. A number of these sites are historic sites that also feature preservation orchards of old-timey apple varieties.

3. Propagation through grafting and propagation through root sprouts are both clonal methods of reproducing apple varieties, whereas propagation by seed will tend to produce unique varieties. One common saying among apple collectors is that 1 in 10,000 apple seedlings will produce a fruit worth eating; the remaining 9,999 apple varieties will be "spitters"—apples so unpleasant in taste that one would spit them out a moment after biting into them.

4. It is important to differentiate here between the root sprouts of seedling trees and the root sprouts of grafted trees. Root sprouts of seedling trees enable the clonal propagation of the unique seedling variety. Root sprouts of grafted trees would only enable the clonal propagation of the rootstock, onto which the unique variety was grafted. Thus, reference to early exchange of varieties via root sprouts is directed at the exchange of the root sprouts of seedling trees.

5. In his 1905 book, Ragan relied on *publications* such as lists from the U.S. Department of Agriculture (USDA), regional agricultural reports, and nursery catalogs. As such, his work favors apple varieties that were exchanged commercially and/or recognized in formal USDA or other agricultural varietal lists. This

figure of 7,000 varieties was calculated by taking a random sample of pages from Ragan's 1904 work, counting the average number of varieties listed on each page, and multiplying the average number by the number of pages listing apple varieties.

6. In the case of finding and grafting seedling trees, the age of the variety can be deduced from estimating the age of the tree.

References

Basso, Keith. 1984. Stalking with stories: Names, places and moral narratives among the western Apache. In *Text, Play, and Story: The Construction and Reconstruction of Self and Society*, edited by E. Bruner, 19–53. Washington, DC: American Ethnological Society.

Beach, S. A. 1905. *The Apples of New York*. Vol. 1. Albany, NY: J. B. Lyon.

Brush, Stephen. 2004. *Farmers' Bounty: Locating Crop Diversity in the Contemporary World*. New Haven, CT: Yale University Press.

Calhoun, Creighton Lee, Jr. 1995. *Old Southern Apples*. Blacksburg, VA: McDonald and Woodward.

Carter, Erica, James Donald, and Judith Squires, eds. 1993. Introduction. In *Space and Place: Theories of Identity and Location*, vii–xv. London: Lawrence and Wishart.

Casey, Edward. 1996. How to get from space to place in a very short stretch of time: Phenomenological prolegomena. In *Senses of Place*, edited by Stephen Feld and Keith Basso, 13–52. Santa Fe: School of American Research Press.

Cruikshank, Julie. 2005. *Do Glaciers Listen? Local Knowledge, Colonial Encounters, and Social Imagination*. Vancouver: UBC Press.

Diamond, David H. 2010. Origins of pioneer apple orchards in the American West: Random seedling versus artisan horticulture. *Agricultural History* 84(4):423–450.

Dove, Michael. 1998. Living rubber, dead land, and persisting systems in Borneo: Indigenous representations of sustainability. *Bijdragen tot de Taal-, Land-en Volkenkunde* 154(1):21–54.

———. 1999. The agronomy of memory and the memory of agronomy: Ritual conservation of archaic cultigens in contemporary farming systems. In *Ethnoecology: Situated Knowledge/Located Lives*, edited by Virginia D. Nazarea, 45–70. Tucson: University of Arizona Press.

Feld, Stephen, and Keith Basso, eds. 1996. Introduction. In *Senses of Place*, 3–11. Santa Fe: School of American Research Press.

Folger, J. C., and S. M. Thomson. 1921. *The Commercial Apple Industry of North America*. New York: Macmillan.

Fowler, Cary, and Pat Mooney. 1990. *Shattering: Food, Politics, and the Loss of Genetic Diversity*. Tucson: University of Arizona Press.

Goland, Carol, and Sarah Bauer. 2004. When the apple falls close to the tree: Local food systems and the preservation of diversity. *Renewable Agriculture and Food Systems* 19(4):228–236.

Harris, Stephen A., Julian P. Robinson, and Barrie E. Juniper. 2002. Genetic clues to the origin of the apple. *Trends in Genetics* 18(8):426–430.

Heald, Paul, and Susannah Chapman. 2012. Veggie tales: Pernicious myths about patents, innovation, and crop diversity in the twentieth century. *University of Illinois Law Review* 4:1051–1102.

Hedrick, U. P. 1950. *A History of Horticulture in America to 1860.* New York: Oxford University Press.

Hensley, Tim. 2005. A curious tale: The apple in North America. In *The Best Apples to Buy and Grow,* edited by Beth Hanson, 4–15. New York: Science Press.

Hmimsa, Y., Y. Aumeeruddy-Thomas, and M. Ater. 2012. Vernacular taxonomy, classification and varietal diversity of fig (*Ficus carica* L.) among Jbala cultivators in northern Morocco. *Human Ecology* 40:301–313.

Holtzman, Jon. 2006. Food and memory. *Annual Review of Anthropology* 35:361–378.

Hunn, Eugene. 1999. The value of subsistence for the future of the world. In *Ethnoecology: Situated Knowledge/Located Lives,* edited by Virginia D. Nazarea, Tucson: University of Arizona Press.

Juniper, Barrie E. 2007. The mysterious origin of the sweet apple. *American Scientist* 95:44–51.

Lape, Fred. 1979. *Apples and Man.* New York: Van Nostrand Reinhold.

Lupton, Deborah. 1994. Food, memory and meaning: The symbolic and social nature of food events. *Sociological Review* 42(4):664–685.

Nabhan, Gary. 1989. *Enduring Seeds: Native American Agriculture and Wild Plant Conservation.* Tucson: University of Arizona Press.

Nazarea, Virginia D. 1998. *Cultural Memory and Biodiversity.* Tucson: University of Arizona Press.

———. 2005. *Heirloom Seeds and Their Keepers: Marginality and Memory in the Conservation of Biological Diversity.* Tucson: University of Arizona Press.

———. 2006. Local knowledge and memory in biodiversity conservation. *Annual Review of Anthropology* 35(17):1–17.

Nora, Pierre, ed. 1996. *Realms of Memory: Rethinking the French Past,* Vol. 1: *Conflicts and Divisions.* Edited by Lawrence D. Kritzman. Translated by Arthur Goldhammer. New York: Columbia University Press.

Peluso, Nancy Lee. 1996. Fruit trees and family trees in an anthropogenic forest: Ethics of access, property zones, and environmental change in Indonesia. *Contemporary Studies in Society and History* 38(3):510–548.

Ragan, W. H. 1905. Nomenclature of the apple: A catalog of the known varieties referred to in American publications from 1804 to 1904. USDA Bureau of Plant Industry Bulletin 56. Washington, DC: U.S. Department of Agriculture.

Richards, Paul. 1996. Culture and community values in the selection and maintenance of African rice. In *Valuing Local Knowledge: Indigenous People and Intellectual Property Rights*, edited by Stephen Brush and Doreen Stabinsky, 209–229. Washington, DC: Island Press.

Rigney, Ann. 2005. Plentitude, scarcity and the circulation of cultural memory. *Journal of European Studies* 35(11):11–28.

Sadiki, M., D. Jarvis, D. Rijal, J. Bajracharya, N. N. Hue, T. C. Camacho-Villa, L. A. Burgos-May, M. Sawadogo, D. Balma, D. Lope, L. Arias, I. Mar, D. Karamura, D. Williams, J. L. Chavez-Servia, B. Sthapit, and V. R. Rao. 2007. Variety names: An entry point to crop genetic diversity and distribution in agroecosystems? In *Managing Biodiversity in Agricultural Ecosystems*, edited by D. I. Jarvis, C. Padoch, and H. D. Cooper, 34–76. New York: Columbia University Press.

Schama, Simon. 1995. *Landscape and Memory*. New York: Alfred A. Knopf.

Soleri, Daniela, and Steven E. Smith. 1999. Conserving folk crop varieties: Different agricultures, different goals. In *Ethnoecology: Situated Knowledge/Located Lives*, edited by Virginia D. Nazarea, 133–154. Tucson: University of Arizona Press.

Veteto, James. 2008. The history and survival of traditional heirloom vegetable varieties in the southern Appalachian Mountains of western North Carolina. *Agriculture and Human Values* 25:121–134.

CHAPTER THREE

Food from the Ancestors

Documentation, Conservation, and Revival of Eastern Cherokee Heirloom Plants

JAMES R. VETETO AND KEVIN WELCH

Southern Appalachia is one of the most biodiverse temperate forest regions in the world (Braun 2001; Cozzo 2004) and has been widely studied by botanists and ecologists (e.g., Martin et al. 1993; Pittillo et al. 1998). A lesser known and studied phenomena is that southern Appalachia has one of the highest currently known levels of agricultural biodiversity in the United States, Canada, and northern Mexico (Veteto 2010). The high levels of forest and agricultural biodiversity in southern Appalachia reinforce the correlation between mountain marginality and biocultural diversity worldwide (Rhoades and Nazarea 1999; Stepp et al. 2005; Rhoades 2007). Recent attempts at reviving the U.S. Biosphere Reserve program in southern Appalachia have recognized agrobiodiversity as a key component in their conservation efforts (Gilbert 2009).

The Eastern Band of Cherokee Indians has the oldest living agricultural tradition in southern Appalachia. The Eastern Cherokee live on approximately 56,000 acres of land in the southwestern part of Appalachian North Carolina and have close to 13,000 active members in the tribe (Perdue 2005; Finger 1991). When first encountered by Europeans in the sixteenth century, the Cherokee were an agricultural people relying heavily on the "three sisters" plant guild of corn (*Zea mays*), beans (predominately *Phaseolus vulgaris* and *P. coccineus*), and squash (*Cucurbita* spp.), supplemented by hunting and gathering a wide diversity of wild foods. Throughout years of exchange with Europeans, they also gradually began to adopt introduced crops such as potatoes (*Solanum tuberosum*), sweet potatoes (*Ipomoea batatas*), cabbage (*Brassica oleracea*), and cowpeas (*Vigna unguiculata*).

The historical pattern of Cherokee agriculture was for men to clear the fields and help with the planting and harvesting and for women to oversee the day-to-day management of the fields (Greene and Robinson 1987). The women worked the fields twice yearly with bone or stone hoes attached to a stick as was prescribed by the Selu corn origin story. Most of the work time was spent protecting the crops from animals such as crows, rabbit, and deer. This task was generally undertaken by older women, who would sit upon high scaffolds overlooking the family gardens to scare wildlife away. Historical Cherokee life was choreographed by an agricultural ceremonial calendar that featured numerous celebrations. For example, the Green Corn Ceremony was a big harvest festival of thanksgiving that featured rituals, feasting, and dancing for several days, as well as the forgiveness of all crimes committed during the previous year, except murder (Greene and Robinson 1987). Today in Eastern Cherokee life, such festivals are still celebrated but on a much smaller scale. Agriculture as a way of life has greatly declined as Cherokee people have sought off-reservation work, tribal government jobs, or work in the tourist industry of Cherokee, North Carolina. Among those Eastern Cherokee who still grow food today, men and women generally work together in small homegardens tended by older generations.

Recent years have seen a revival of interest in Cherokee gardening and heirloom seeds, and we focus here on several aspects of this ongoing trend. First, we provide a detailed overview of existing Eastern Cherokee agrobiodiversity and examine farmer decision making related to the resilience of heirloom cultivars. Second, we discuss the use of Cherokee agrobiodiversity in tribal culinary practices as a prominent example of how culturally salient traditions promote the continued *in vivo* use and circulation of Cherokee heirloom seeds. Finally, we describe two tribal institutions, the Cherokee Indian Fair Agricultural Exhibit and the Center for Cherokee Plants, as examples of how *in situ* conservation is accomplished by emphasizing culturally salient motifs that encourage the continuation of threatened agricultural lifeways *in vivo* among the tribe.

Eastern Cherokee Agrobiodiversity and Farmer Decision Making

In the fall of 2008 we jointly conducted a study in collaboration with the Center for Cherokee Plants to investigate the survival of Eastern Cherokee heirloom food plants. Participants in the study were recruited by the Center for Cherokee Plants from growers they had worked with previously, and a chain-referral sampling methodology was used to identify and recruit additional participants. In all, fifteen Cherokee heirloom gardeners participated. In-depth oral history interviews were conducted to document biological and cultural aspects of Cherokee heirloom food plants and to investigate the underlying motivations that growers have for continuing to grow traditional heirloom varieties. A benchmark socioeconomic survey was also administered to understand how Eastern Cherokee gardeners were distributed according to variables such as age, gender, education, income level, and spiritual beliefs. The methodologies generally followed those established by Virginia Nazarea (2006) for the "memory banking" of farmers' cultural and agroecological knowledge about traditional cultivars to complement the more traditional scientific *ex situ* conservation strategy of collecting and storing folk crop varieties in seed bank facilities (also see chapter 1, this volume).

Although fifteen is a small number of informants to interview, we are confident that the interviews provided us with a sufficient grasp of contemporary Cherokee agrobiodiversity. Coauthor Kevin Welch, a native of the traditionalist Big Cove Community, knows almost everyone in the tribe and had been seeking out growers of Eastern Cherokee heirloom vegetable varieties for the previous five years, but not very many Eastern Cherokee gardeners and farmers still grow heirloom varieties. The results from the socioeconomic surveys (see table 3.1) showed that the Eastern Cherokee heirloom growers whom we interviewed were predominately male, elderly, retired, and low income. On average, each grower was maintaining about twelve heirloom varieties on 1.68 acres of land, and all but two of the growers were producing primarily for home consumption in small homegardens. These elderly growers were mostly Baptist, and they had achieved, on average, an eleventh grade education.

Despite the relatively low number of growers, the agrobiodiversity they are still maintaining is relatively high: 32 species and 128 distinct folk crop varieties are still being grown by Eastern Cherokee gardeners (see table 3.2). Beans were most numerous among heirloom cultivars (n = 45), followed by apples (*Malus pumila*, n = 20) and corn (n = 14). Some of the most culturally significant varieties include Cherokee White Flour corn, Cherokee Tender October beans (*P. vulgaris*), Cherokee butterbeans

Table 3.1. Socioeconomic data from 15 Eastern Cherokee heirloom growers

Category	Average	Total
Number of heirloom varieties grown	12.33	185*
Acres in production	1.68	25.18
Gender		14 male, 1 female
Age	70.07	
Annual household income (n=11)**	$25,000 (median)	
Occupation		Various, mostly retired (n=8)
Years of education (n=13)**	10.69	
Religion/spirituality		Various, mostly Baptist (n=10)

*Includes fifty-six varieties that are maintained by multiple growers, so the total is higher than the 128 total distinct heirloom varieties grown cited elsewhere in this study.
**Several growers chose not to provide this data on the survey, and this is reflected in the lower number of responses.

Table 3.2. Eastern Cherokee folk crop varieties documented in the present study

Plant Type	Scientific Name	Total Varieties (N=128)	Examples of Local Variety Names
Bean	*Phaseolus vulgaris* *P. coccineus* *Dolichos lablab* *Richinus communis*	45	Tender October, Cherokee Butterbean, Yellow Hull Cornfield, Greasy Cutshort, Striped Creaseback
Apple	*Malus pumila*	20	Green-Striped Winesap, Horse, Lunsford, Sheepnose, Stamen
Corn	*Zea mays*	14	Cherokee White Flour, Cherokee Yellow Flour, Pearl Hominy
Greens	*Brassica* spp. *Lepidum sativum* *Rorippa nasturtium*	8	Cherokee mustard, Creasy Greens, watercress, Winter mustard

Table 3.2. Eastern Cherokee folk crop varieties documented in the present study (continued)

Plant Type	Scientific Name	Total Varieties (N=128)	Examples of Local Variety Names
Squash/pumpkin	*Cucurbita maxima* *C. argyrosperma* *C. pepo*	8	Old-time Pie pumpkin, Roughbark Cherokee Candy Roaster, Cushaw, White Winter squash
Tomato	*Lycopersicon esculentum*	7	Cherokee Purple, Walter Johnson Stripey
Gourd	*Lagenaria siceraria* *Luffa acutangula* *Trichosanthes anguina*	6	Caveman, Vine Okra, Dipper, Snake
Okra	*Abelmoschus esculentus*	3	Red, Green
Cowpea	*Vigna unguiculata*	3	Clay, Little Red Field, Whippoorwill
Grape	*Vitis* spp.	2	Pink, Purple
Peach	*Prunus persica*	2	Purple Indian, White Indian
Cherry	*Prunus avium*	1	Wild
Gooseberry	*Ribes* spp.	1	Gooseberry
Ground cherry	*Physalis pubescens*	1	Yellow
Jerusalem artichoke	*Helianthus tuberosus*	1	Jerusalem Artichoke
Jobs tears	*Coix lacryma-jobi* var. lacryma-jobi	1	Cornbeads
Peanut	*Arachis hypogaea*	1	Georgia Red
Pear	*Pyrus communis*	1	Barlett
Plum	*Prunus* spp.	1	Wild
Potato	*Solanum tuberosum*	1	Irish Cobbler
Rhubarb	*Rheum rhabarbarum*	1	Rhubarb

FIGURE 3.1. Cherokee butterbeans (*Phaseolus coccineus*) displayed in a traditional Cherokee basket at the Cherokee Indian Fair Agricultural Exhibit. A basket of popcorn (*Zea mays*) is in the background. Photo by Keith Nicholson, used with permission.

(*P. coccineus*; see figure 3.1), Cornbeads (*Coix lacryma-job* var. lacryma-jobi), Old-Time Cherokee mustard (*Brassica juncea*), Irish Cobbler potato, Yellow ground cherry (*Physalis pubescens*), Sheepnose apple, White Indian peach (*Prunus persica*), and Cherokee Roughbark Candy Roaster squash (*Cucurbita maxima*).

Motivations for Seedsaving and Heirloom Gardening

To investigate motivations for seedsaving, the informants were simply asked, "What makes this a variety that you like to grow?" Varieties that they cited in a free list activity, along with their reasons, were recorded without further prompting. This slight turn-of-words from the "What is this plant used for?" that is more often employed in ethnobotanical studies was used in an attempt to provide for a wider spectrum of farmer motivations for seedsaving. The reasons that Eastern Cherokee growers gave for continuing to grow and maintain heirloom cultivars were numerous and varied (see table 3.3). Grower responses were coded and empirically grouped into categories. Cultural, ecological, economic, and other reasons for persistence were given in various ratios. We later organized responses further into two broad categories, cultural and utilitarian importance or salience. Some of the responses that Cherokee growers gave could be grouped into neither category; twelve responses fit into this "other" category, including maintaining heirlooms for curiosity, sharing, good smell, and unique appearance.

We use the "cultural" versus "utilitarian" categories in the analysis that follows, although it is probably ultimately the case that categories proposed by scientific researchers to organize and structure cultural phenomena are too interconnected to be considered separate domains in holistic cultural systems. Cognitive ecological anthropologists such as Gregory Bateson have previously pointed this out: "Our categories 'religious,' 'economic,' etc. are not *real* subdivisions which are present in the cultures we study, but are merely *abstractions* which we make for our own convenience when we set out to describe cultures in words. They are not phenomena present in culture, but are labels for various points of view which we adopt in our studies" (1972:64). Tim Ingold (2000) points out that such categorizing behavior by scientists has its roots in the Western natural philosophy tradition (e.g., Aristotle, Descartes, Kant) based on a strict dichotomy between mind and nature and between nature and culture.

Cultural and utilitarian reasons for decision making cannot be completely separated. For example, culinary traditions and tastes (ethnogastronomy) are highly cultural in nature, but they also fulfill very practical nutritional needs for humans. Yet, despite the role of food in providing basic biological sustenance for human survival, it is still the case that nutritional needs can be met in a diversity of ways and that resulting gastronomic traditions are heavily tempered by cultural norms. As Paul Minnis has pointed out, "A meal of peanut butter and fried worms topped off with chiles may be very nutritious, but it would be unacceptable to most North Americans because the foods are combined in culturally, not biologically, inappropriate ways" (2000:3). The reasons for agrobiodiversity persistence that guide farmer decision making and their relationship to food traditions can be more properly understood as being intermingled or biocultural in nature (Maffi 2001; Veteto and Skarbø 2009).

Table 3.3. Grower motivations for maintaining heirloom varieties*

Reason	Number of responses
Cultural salience	
Specific culinary preferences	211
Taste/flavor	90
Cultural heritage	51
Display/compete at Cherokee Indian Fair	26
Aesthetics	15
Cherokee jewelry making	4
Sharing with others	4
Cultural education	3
Total	404
% relative to utilitarian salience	83.64
Utilitarian salience	
Food preservation quality	31
Market value	16
Vegetable quality	8
Animal feed	6
Size	4
High yielding	3
Water container	2
To increase seed stock	2
Local adaptation	1
Disease resistance	1
Fast cooking	1
Pest control	1
Easy to grow	1
Cover crop	1
For bean trellis	1
Total	79
% relative to cultural salience	16.35

*Individual responses were coded from in-depth oral history interviews with fifteen growers. For each heirloom cultivar the grower was asked, "What makes this a variety you like to grow?"

The more traditional categories of agronomic, economic, and ecological reasons for farmer decision making and agrobiodiversity persistence are not easily separated either, despite the tendency of previous researchers to rarely question their validity and accuracy (e.g., Bellon 1991). For example, if a farmer chooses to grow a folk crop variety for the agronomic reason that it produces well, that characteristic is most likely also related to its adaptation to local ecological conditions and may also be related to its success at local markets. Thus, categorizing reasons for farmer decision making is largely an heuristic device for researchers to better explain what is influencing farmer decision making. We need to keep in mind that such decisions are rarely made along variables that are completely isolated from other influencing factors. Actual decision making is also heavily tempered by a preattentive process that is holistic and not easily broken down into categories: "The preattentive process is a nondeliberate simplification that hinges on the actor's 'feel' of the situation. It narrows down the range of alternatives from those possible to those feasible and thereby sets the stage for deliberate or attentive consideration of the remaining options" (Nazarea-Sandoval 1995:16).

Despite the complexity of agricultural decision making, coding participant responses and categorizing them are useful exercises for helping researchers understand why farmers make decisions. In categorizing reasons for agrobiodiversity persistence among the Eastern Cherokee as being either cultural or utilitarian in nature (or "other" if they are largely idiosyncratic and do not qualify for either category), we fully recognize the need to conduct more research into investigating the cultural context of agrobiodiversity research (Brush 1992; Perales et al. 2005). By contrasting cultural salience with utilitarian salience (agronomy, economy, ecology) in interpreting agrobiodiversity persistence, a clearer picture of why growers still choose to maintain high levels of agrobiodiversity in a postagrarian area of the most industrialized nation in the world emerged. The results of this study indicate that cultural salience is the prevailing motivation for Eastern Cherokee growers to perpetuate heirloom cultivars.

Culinary Traditions

The Eastern Cherokee have a unique and varied culinary tradition that includes many traditional dishes that are prepared from heirloom varieties. We include a brief overview of the most popular Eastern Cherokee dishes so the reader can get a sense of how Cherokee agrobiodiversity is transformed into culturally valued foodstuffs through the medium of cooking and other food preparation technologies. Bean bread (Cherokee: *Tu-ya-di-su-yi-ga-du*) is a mixture of cornmeal, flour, and cooked beans that are mixed together, wrapped with soaked hickory (*Carya* spp.) leaves and tied together with young river grass, and low-boiled for about thirty minutes (Plemmons et al. 2000). The bean bread is unwrapped after cooking and can be eaten with toppings such as animal grease or cooked greens. Cherokee heirloom corn and bean varieties are favored by traditional cooks for making bean bread and often include Cherokee White Flour corn and Cherokee butterbeans as the main ingredients.

Unlike bean bread, leather breeches is a food preservation and culinary tradition that white Appalachian settlers adopted from the Cherokee and is still used widely by old-time Appalachian gardeners and farmers. Leather breeches (Cherokee: *A-ni-ka-yo-su-hi-tu-ya*) are green beans that are prepared by a traditional method of picking them when they make a full bean in the pod, taking a needle and thread and stringing dozens of bean pods together, and then hanging them up in a dry area to save for winter cooking. A more modern way of preparation is to cut the pods in half and remove the strings, lay them out in an area such as a greenhouse, and then store them in zip-top plastic bags until they are needed for cooking. In wintertime, the leather breeches are soaked overnight in lightly salted water and then cooked for several hours with fatback or pieces of bacon and a little bit of salt. The taste of leather breeches is unique and quite different from green beans that are cooked fresh. Preferred heirloom varieties for making leather breeches include Cherokee October beans, Yellow Hull Cornfield beans, Greasyback beans, and White Half-runner beans (all *P. vulgaris*).

Corn, Beans, and Walnut (Cherokee: *Ce-di Selu I-asa Asu-yi*) is a mixture of corn that has been processed into hominy and cooked beans that is and flavored with a paste made from black walnuts (*Juglans nigra*). This dish is typically sweetened with honey or sugar and eaten as a dessert. Heirloom corn and bean varieties such as Cherokee White Flour corn, White Hickory King corn, Cherokee October beans, or Cherokee Cornfield beans (*P. vulgaris*) are preferred to give the dish its desired flavor. Other dishes such as Candy Roaster and Cushaw Fritters (Cherokee: *U-ja-she-gwa U Je-sdi*), Sweet Potato Bread (Cherokee: *Oo-gu Na Sti-nu-nv Ga-du*), Gritted Bread, Hickory Nut Soup (Cherokee: *Ga Na-sti*), Persimmon Pudding (Cherokee: *Sa-li*), and Hominy Corn Drink (Cherokee: *Gu-no-he-nv*) complement these "anchor" dishes and utilize heirloom garden plants and wild harvested foods in an extremely diverse Cherokee culinary repertoire (Plemmons et al. 2000).

Our research results indicate that almost all Cherokee heirloom food plants are grown because of the flavor they impart into traditional Cherokee dishes. That being the case, efforts at preserving Cherokee cuisine such as the publication of Cherokee cookbooks (e.g., Gwaltney 1988; Plemmons et al. 2000), serving traditional Cherokee foods at the Cherokee Indian Fair, and hosting community potlucks featuring traditional dishes by local organizations such as the Center for Cherokee Plants are valid and potent strategies for promoting the conservation of Cherokee agrobiodiversity.

The Cherokee Indian Fair Agricultural Exhibit

The Cherokee Indian Fair is a fall festival that has been ongoing since 1914. It is held every October, and Joan Greene and H. F. Robinson (1987) have argued that it represents a modified carryover of the Green Corn Ceremony, which was traditionally held near the end of September when the corn crop had matured. The agricultural exhibit at the fair gives Cherokee growers a chance to compete at growing traditional Cherokee crops. From our observations, the agricultural exhibit gives Cherokee growers important incentives for promoting agrobiodiversity conservation and agricultural innovation using traditional crops. Major awards are given in categories of traditional Cherokee food crops and plants. In addition, local plant breeders are sometimes rewarded for showy innovations they have made on traditional cultivars. In the fall of 2008, when coauthor James Veteto had the honor of being one of the judges at the agricultural exhibit, the competing categories for the agricultural exhibit were numerous (listed in table 3.4). These prize categories encourage people to grow a wide diversity of traditional Cherokee plants. Many of the growers that we interviewed participated in the agricultural exhibit, and for several it was the main venue and motivation for continuing to grow out traditional varieties (figure 3.2).

A second theme that the Cherokee Indian Fair Agricultural Exhibit promotes, though perhaps unintentionally, is that of encouraging the continuation of Cherokee agricultural innovation. The Cherokee have nurtured their heirloom varieties for hundreds—if not thousands—of years and have carefully selected cultivars that are adapted to local soil types and resistant to pests and diseases, in addition to tasting good in traditional culinary dishes and playing an important role in cultural history and identity. However, since most Cherokee today do not depend on agriculture for their subsistence, it is likely that agricultural innovation through selection and adaptation to a changing environment has slowed considerably. During the course of oral history interviews, it became apparent that the agricultural exhibit provides a more modern venue for continuing cultivation, selection, and innovation of Eastern Cherokee crops. Two examples of local crop breeding illustrate this point. One Cherokee gardener had a lavender variety of Cherokee October bean in his collection, a rare and unique color that is not often seen in October beans in western North Carolina. When asked about this "anomaly," the grower replied:

> They had two or three colors of October beans [previously] and the other lavender bean was a butterbean. And seeing Dad develop stuff I said, "I wonder if I could get that color in these other beans?" So

Table 3.4. Competition categories at the Cherokee Indian Fair Agricultural Exhibit

Corn multicolored kernels (*Zea mays*—thirty ears)

Indian Flour corn (*Zea mays*—thirty ears, no dent, eight rows of kernels per ear—yellow, white, other colors)

Indian beans (*Phaseolus vulgaris*, *P. coccineus*—one peck, shelled and dried, displayed in an Indian basket, including October beans and butter-beans and other traditional Cherokee bean types)

Corn beads (*Coix lacryma-jobi* var. *lacryma-jobi*—1/2 gallon, displayed in an Indian basket)

Traditional crops of the Cherokee (a display of three to five different traditionally cultivated crops, including the Cherokee and English names of each crop)

Field corn (*Zea mays*—ten ears; white, yellow, and other colors)

Honey production (three jars—without comb, with comb)

Sweet potatoes (*Ipomoea batatas*—one peck)

Winter squash (*Cucurbita* spp.)

Candy Roaster (*Cucurbita maxima*)

Cushaw (*Cucurbita argyrosperma*, orange or green striped)

Irish potatoes (*Solanum tuberosum*—one peck, white and red)

Other winter squash (*Cucurbita* spp.)

Pumpkin (*Cucurbita* spp.—largest, ugliest, painted)

Ornamental gourds (*Lagenaria siceraria, Trichosanthes anguina*; fresh, undecorated, displayed in an Indian basket)

Other pumpkin (*Cucurbita* spp.)

Herb display (five different fresh or dried herb plants, all labeled, with Cherokee name and plant uses, including food, flavor, medicine, dye, or fiber)

Any other traditional Cherokee crop (wild or cultivated, must include a card with an explanation of what it is and how it is used, and the Cherokee name and plant uses)

Largest gourd (*Lagenaria siceraria*)

Largest sunflower (*Helianthus annuus*, diameter of head)

Unusual vegetable

Largest Candy Roaster (*Cucurbita maxima*)

Popcorn (*Zea mays*—five ears, displayed in an Indian basket)

FIGURE 3.2. Kevin Welch and Sarah McClellan-Welch attaching award ribbons at the Cherokee Indian Fair Agricultural Exhibit. They are holding a snake gourd (*Trichosanthes anguina*). Photo by Keith Nicholson, used with permission.

I planted them together and it was three or four years and I was going through—well I'll never forget it, up in the holler—and there was a pod shaped like a butterbean but it was much smaller. I knew it wasn't a butterbean. I said, "I wonder," so I marked it and it got dry shelly and it was the same shape as a butterbean, but it wasn't that big. It had the color I was looking for. So the next year I began to get the color in the October beans and in—I call the kidney-type beans—I got color in them too. [This is interesting since mountain butterbeans and October/kidney beans are two different species, *P. coccineus* and *P. vulgaris*.] (Veteto 2010:105)

This Cherokee gardener is a local legend for winning many of the grower categories at the Cherokee Indian Fair and has done so for many years. The Agricultural Exhibit provides him motivation for breeding new variation into his Cherokee heirloom seed stock, and he is rewarded for his agricultural innovations by winning at the fair. However, his success has not precluded continued propagation and maintenance of his other distinct, traditional Cherokee cultivars, as he grows out and displays them as well. The overall diversity in his collection of seed stock is being increased through his informal plant breeding without sacrificing the original germplasm from which it originated.

Another Eastern Cherokee grower has developed a multicolored Cherokee Flour corn that has white, yellow, blue, purple, and red kernels. This colorful corn variety is in contrast to the white and yellow flour corn varieties, which are widely acknowledged as the long-time cultivars of the tribe. He has selected and bred it to display at the Cherokee Indian Fair Agricultural Exhibit, just as the grower described above has been doing with October beans. According to him:

> I did it on purpose to get the kernels bigger on the Indian corn. What it does is make a wider grain.
>
> [I have been breeding] these, maybe fifteen years. Probably been going on longer than that. . . . There was a big competition at the festival. Each family would try to win it a long time ago. It's not that way anymore, not as much as it used to be. They would just about fight over first prize at the festival.
>
> [I cross] mostly just old flour corn really. Sometimes it gets mixed in with field corn but that depends on when you plant it and where you plant it . . . The old flour corn here is the white that you see in it. You can take all this out and plant it and it will eventually turn all back to white and take the color out of it. I usually plant a field of white but this year I didn't and they made a yellow. This is a yellow, it's just like the white and it will be this color [gestures]. So, this is all mixed up, it's really white and yellow [flour corns]. (Veteto 2010:108)

Note that the agricultural exhibit was providing motivation for Cherokee grower innovation, but for aesthetic reasons that are likely different from motivations of Cherokee farmers hundreds or thousands of years ago. However, even though these newer Cherokee varieties that are being developed by growers from traditional seed stock to display at the agricultural exhibit are being grown for aesthetic reasons, they are at the same time being adapted to local environmental conditions as they are bred and grown out in contemporary Cherokee gardens. And again, this grower is creating new varieties out of old seed stock but is also careful to keep the original seed stock pure from mixing and being lost.

The Cherokee Indian Fair Agricultural Exhibit provides a venue for continuing Cherokee agricultural innovation in a more modern setting and helps ensure that Cherokee agrobiodiversity continues to evolve while also saving time-honored and cherished heirloom varieties. The agricultural exhibit acts to promote Cherokee agricultural and wild plant diversity by providing a community outlet for celebrating Cherokee cultural identity and traditional plant use. The inclusion of the agricultural exhibit in a larger cultural event such as the Cherokee Indian Fair provides a performative, and edible, link between culture and agriculture that perpetuates the cultural relevance of Cherokee agrobiodiversity. Although the agricultural exhibit is not an official or intentional conservation program, it directly engages several of Nazarea's (2005) suggestions for supporting and promoting *in vivo* agricultural lifeways that are the cornerstone for any agrobiodiversity conservation efforts. The fair, by recognizing Eastern Cherokee gardeners as creators and curators of agrobiodiversity and by giving incentives for seedsavers to propagate Cherokee heirloom cultivars, serves as a tangible site of memory and indirect promoter of *in situ* conservation. It also engages the milieu of memory of local gardeners as they maintain local varieties in their fields to be able to compete in the fair. Memories of past agricultural exhibit events are also kept alive in the circulation of local Cherokee stories (Veteto 2010; see also chapter 2, this volume).

The Center for Cherokee Plants

The Center for Cherokee Plants is a conservation program that was officially established by coauthor Kevin Welch in 2007, and work on the center's projects has been ongoing since 2005. The motto of the center is "Putting Culture back into Agriculture" (see figure 3.3). It is located at the traditional Kituwah "mothertown" sacred site on two acres of land that contain several abandoned dairy buildings that are being remodeled for the center's use. The land was donated by the business committee of the Eastern Band of Cherokee Indians (McClellan-Welch 2008). Welch, an enrolled member of the Eastern Band of Cherokee Indians and born and raised in the traditionalist Big Cove Community, spent many years away from the reservation working at different professions. Upon returning in 2000, he began to search around for the old-time Cherokee cultivars that he remembered from his youth, such as Cherokee October beans and Rattlesnake pole beans. He found that far fewer Cherokee people were growing out traditional cultivars than in the past and that the growers were elderly and spread out in small pockets across different Cherokee communities. Many of the growers possessed seed stocks that were so low that they could no longer share seeds with their neighbors, a time-honored Cherokee tradition. Welch was disturbed by

FIGURE 3.3. Official logo of the Center for Cherokee Plants. Its motto is "Putting Culture back into Agriculture." Photo by James R. Veteto.

the limited availability of traditional Cherokee seeds and plants, so he started the Center for Cherokee Plants as a way to conserve, promote, and revitalize Cherokee seeds, plants, and foodways.

Since 2007, the Center for Cherokee Plants has been engaging in grow-outs of Cherokee heirloom seeds and making them available to the local community. The center participates in the Chief's Cherokee Family Garden Project to help get heirloom seeds back in the hands of Cherokee growers and to promote local gardening and food production. In addition to seed conservation and distribution, the center has also established a tribal plant nursery to grow out plants that are utilized by Cherokee artists, along with wild food plants, medicinal plants, wildlife habitat and erosion control plants, and heirloom fruit varieties. The nursery also serves as a repository for plants that have been rescued from local construction sites (McClellan-Welch 2008). The Center for Cherokee Plants sponsors educational programs on traditional Cherokee agriculture throughout southern Appalachia and has engaged in outreach, networking, and consultation with heirloom seed conservation projects of several other American Indian tribes. The center also periodically hosts potlucks highlighting traditional Cherokee foods, which provides a venue for dishes cooked with heirloom varieties to be appreciated by the larger Cherokee community.

Promoting Cherokee culture is central to the mission of the Center for Cherokee Plants. Their cultural approach to conservation has a high degree of success and is consistent with the findings of this research—that local growers are maintaining folk crop varieties largely because of cultural relevance. Conservation initiatives among other indigenous people, for example, "cultures of the seed" in the Peruvian Andes (Gonzales 2000; see chapters 4 and 5, this volume), have also been successful by promoting the conservation of biological and cultural diversity through cultural themes (Nazarea 2006).

Conclusions

Agrobiodiversity conservation programs worldwide have seen a trend toward *in situ* strategies over the past twenty-five years (Maxted et al. 1997; Brush 2000; Hammer 2003). Although recognized as being complementary and in some ways superior to *ex situ* strategies, programmatic *in situ* conservation initiatives have been critiqued for their shortcomings (see, e.g., Nazarea 2005). Nonetheless, community-based conservation efforts such as those undertaken by the Eastern Cherokee have a lot of potential for conserving agrobiodiversity and at the same time celebrating and strengthening cultural identity and *in vivo* agricultural lifeways. Indeed, most seedsaving among the Eastern Cherokee, in the Mountain South in general (Veteto 2010), and among agriculturalists worldwide (Nazarea 2005) is carried on *in vivo* and has been largely beyond the reach of conservation initiatives. Any successful *in situ* conservation effort must therefore have direct relevance to the everyday practices of farmers and gardeners in the communities in which they are operating, since such individuals are both the originators and perpetuators of diverse cultivars in the gardens and fields of the world.

From the long-running Cherokee Indian Fair Agricultural Exhibit to the more recent advent of the Center for Cherokee Plants, the Eastern Cherokee have had community mechanisms in place to conserve agrobiodiversity. These strategies have become more important as, for various reasons, the number of Eastern Cherokee practicing traditional agriculture decreased dramatically in the last forty years. From a plant conservation and historical perspective, Cherokee efforts at *in situ* conservation are extremely important as the Eastern Cherokee are the original agriculturists of southern Appalachia—the region with the highest known levels of agrobiodiversity in much of North America—and progenitors of much of southern Appalachian agrobiodiversity.

Our research indicates that Cherokee heirloom growers are continuing to maintain their folk crop varieties for reasons that are largely cultural in nature. Eastern Cherokee culinary traditions are strongly linked to Cherokee culture and identity. Although generally in decline due to the spread of modern American foods, these traditions have been preserved to a large extent through family customs, community gatherings and celebrations, and the local publication of Cherokee cookbooks. The Cherokee Indian Fair Agricultural Exhibit is linked to traditional Cherokee life patterns through its proximity to the time of the Green Corn Ceremony and provides a venue for Cherokee growers to exhibit and take pride in their Cherokee identity as an agricultural people. It is as much a cultural as it is an agricultural event. The successes that the Eastern Cherokee have had conserving their highly diverse agrobiodiversity repertoire over time, and the interest that present-day programs focusing on celebrating and reviving Cherokee culture and agriculture have generated, provide a compelling rationale for incorporating cultural heritage programs into existing *in situ* conservation efforts.

References

Bateson, Gregory. 1972. Culture contact and schismogenesis. In *Steps to an Ecology of Mind*, edited by G. Bateson, 61–87. New York: Ballantine Books.

Bellon, Mauricio R. 1991. The ethnoecology of maize variety management: A case study from Mexico. *Human Ecology* 19(3):389–418.

Braun, E. L. 2001 [1950]. *Deciduous Forests of Eastern North America*. 2nd ed. Caldwell, NJ: Blackburn Press.

Brush, Stephen B. 1992. Ethnoecology, biodiversity and modernization in Andean potato agriculture. *Journal of Ethnobiology* 12(2):161–185.

———, ed. 2000. *Genes in the Field: On-Farm Conservation of Crop Diversity*. Rome: International Development Research Centre (Canada); International Plant Genetic Resources Institute.

Cozzo, David N. 2004. Ethnobotanical classification system and medical ethnobotany of the Eastern Band of the Cherokee Indians. Ph.D. Dissertation, Department of Anthropology, University of Georgia, Athens.

Finger, John R. 1991. *Cherokee Americans: The Eastern Band of Cherokee Indians in the Twentieth Century*. Lincoln: University of Nebraska Press.

Gilbert, V. C. 2009. Biosphere reserves and food project summary. Unpublished MS proposal.

Gonzales, Tirso A. 2000. The cultures of the seed in the Peruvian Andes. In *Genes in the Field: On-farm Conservation of Crop Diversity*, edited by Stephen Brush, 193–216. Rome: International Development Research Centre; International Plant Genetic Resources Institute.

Greene, J., and H. F. Robinson. 1987. Maize was our life: A history of Cherokee corn. *Journal of Cherokee Studies* 11:40–52.

Gwaltney, F. 1988. *Corn Recipes from the Indians*. Cherokee, NC: Cherokee Publications.

Hammer, Karl. 2003. A paradigm shift in the discipline of plant genetic resources. *Genetic Resources and Crop Evolution* 50:3–10.

Ingold, Tim. 2000. *The Perception of the Environment: Essays in Livelihood, Dwelling, and Skill*. London: Routledge.

Maffi, Luisa, ed. 2001. *On Biocultural Diversity: Linking Language, Knowledge, and the Environment*. Washington, DC: Smithsonian Institution Press.

Martin, W. H., S. G. Boyce and A. C. Echternacht, eds. 1993. *Biodiversity of the Southeastern United States: Upland Terrestrial Communities*. New York: John Wiley.

Maxted, Nigel, Brian Ford-Lloyd, and J. G. Hawkes, eds. 1997. *Plant Genetic Conservation: The In Situ Approach*. London: Chapman and Hall.

McClellan-Welch, Sarah. 2008. Agriculture exhibits for the 2008 fair. *Cherokee One Feather*, October.

Minnis, Paul E. 2000. Introduction. In *Ethnobotany: A Reader*, edited by Paul E. Minnis, 3–10. Norman: University of Oklahoma Press.

Nazarea, Virginia D. 2005. *Heirloom Seeds and Their Keepers*. Tucson: University of Arizona Press.

———. 2006 [1998]. *Cultural Memory and Biodiversity*. Tucson: University of Arizona Press.

Nazarea-Sandoval, Virginia D. 1995. *Local Knowledge and Agricultural Decision Making in the Philippines: Class, Gender, and Resistance*. Ithaca, NY: Cornell University Press.

Perales, Hugo R., Bruce F. Benz, and Stephen B. Brush. 2005. Maize diversity and ethnolinguistic diversity in Chiapas, Mexico. *Proceedings of the National Academy of Sciences of the USA* 102:949–954.

Perdue, Theda. 2005. *The Cherokees*. New York: Chelsea House.

Pittillo, J. Dan, Robert D. Hatcher, and Stanley W. Buol. 1998. Introduction to the environment and vegetation of the southern Blue Ridge Province. *Castanea* 63:202–216.

Plemmons, N., T. Plemmons, and W. Thomas. 2000. *Cherokee Cooking: From the Mountains and the Gardens to the Table*. Gainesville, GA: Georgia Design and Graphics.

Rhoades, Robert E. 2007. *Listening to the Mountains*. Iowa: Kendall/Hunt.

Rhoades, Robert E., and Virginia D. Nazarea. 1999. Local management of biodiversity in traditional agroecosystems: A neglected resource. In *Biodiversity in Agroecosystems*, edited by W. W. Collins and C. O. Qualset, 215–236. Boca Raton, FL: Lewis/CRC Press.

Stepp, John Richard, Hector Castaneda, and Sarah Cervone. 2005. Mountains and biocultural diversity. *Mountain Research and Development* 25(3): 223–227.

Veteto, James R. 2010. Seeds of persistence: Agrobiodiversity, culture, and conservation in the American Mountain South. Ph.D. Dissertation, Department of Anthropology, University of Georgia, Athens.

Veteto, James R., and Kristine Skarbø. 2009. Sowing the seeds: Anthropological contributions to agrobiodiversity studies. *Culture and Agriculture* 31(2):73–87.

CHAPTER FOUR

Sense of Place and Indigenous People's Biodiversity Conservation in the Americas

TIRSO GONZALES

Place for indigenous peoples is where language, culture, daily life, spiritual ceremonies, and rituals nest and dynamically interact. Not all indigenous peoples are agriculturalists; however, for most of them, life revolves around agriculture. This is the case for Andean indigenous peoples of Colombia, Ecuador, Peru, and Bolivia. Its total population is around 17 million individuals, and a significant segment of this population still has strong ties to the land and a unique complement of culture- and place-based strategies of agrobiodiversity conservation. Between the indigenous and the Western Euro-American centered, there are two different ways of knowing (epistemologies), of being (ontologies), and of relating to life and the cosmos (Nakashima et al. 2012; Pimbert 1994a, 1994b; Posey 1999). For the last seventy years, through different top-down strategies and paradigms, rural and agricultural development has been, in many significant ways, eroding indigenous peoples' places in the Andes, and the Americas as a whole (Escobar 1995; Gonzales and Gonzalez 2010; IAASTD 2009; LaDuke 1990; Rengifo 2010; Tauli-Corpuz et al. 2010).

Conceptual Framework

Different worldviews imply different ways of accessing knowledge and thus require different epistemologies (Nakashima et al. 2012; Posey 1999).[1] Conservation *in situ* implies different stakeholders, goals, and cognitive bases. For indigenous and peasant communities, their local traditional knowledge system is the primordial component for biodiversity conservation and regeneration, ecosystem management, and food security. Three major associated explanatory concepts are relevant for this research. First, Eurocentrism theoretically "postulates the superiority of Europeans over non-Europeans" (Battiste and Henderson 2000:21).[2] Second, "coloniality of power," manifests as economic exploitation and race classification (Quijano 2000). Third, "coloniality of knowledge" constitutes knowledge and truth as generated by science (Lander 2000). These concepts allow us to deconstruct and analyze current national and international assessments of food security, agriculture, ecosystems, sustainability, and biodiversity. They also facilitate the deconstruction of the colonial past and present, their associated cognitive systems, development paradigms, and worldviews, as well as their dominant presence today in the Americas. This awareness is illustrated by Latin American society, politics, culture, and related institutions, in particular, at the nation-state level, as "within the past four decades, as the international community codified its norms regarding human and civil rights, Latin American states began moving away from their politics of exclusion regarding indigenous peoples, and, in varying degrees, towards a new politics of inclusion" (Chase 2003:46; see also Garavito 2011; Iturralde and Krotz 1996; Stavenhagen 1990, 2002; Varese 1996, 2001).

I argue that the scientific concept of *in situ* conservation proposed by the science of conservation biology in the 1970s emanates from a fragmented Eurocentric perspective/worldview/paradigm (see figure 4.1). *In situ* conservation was proposed to address the limitations found in the scientific strategy of *ex situ* conservation. However, both scientific strategies obtain only partial results, as they have left out not only indigenous people's worldviews and their sound, holistic strategies of *in situ* conservation, but also issues central to the constitutive elements of the indigenous agricultural knowledge system (Escobar 1998, 1999; Gonzales et al. 2010; IAASTD 2009; Sain and Calvo 2009) and the long-ignored indigenous peoples' agenda of autonomy and self-determination, control over lands and territories, and food and seed security (Declaration of Atitlán 2002; Declaration of Nyéléni 2007; Ishizawa 2003; IPAF 2010; Indigenous Environmental Network 2001; Pajares 2004; Pimbert 2006; PRATEC 2006a, b; Rengifo 2001, 2003, 2004; Santa Fe Board of County Commisioners 2007; Tauli-Corpuz et al. 2010; Valladolid 2005; Valladolid 2001, 2005).

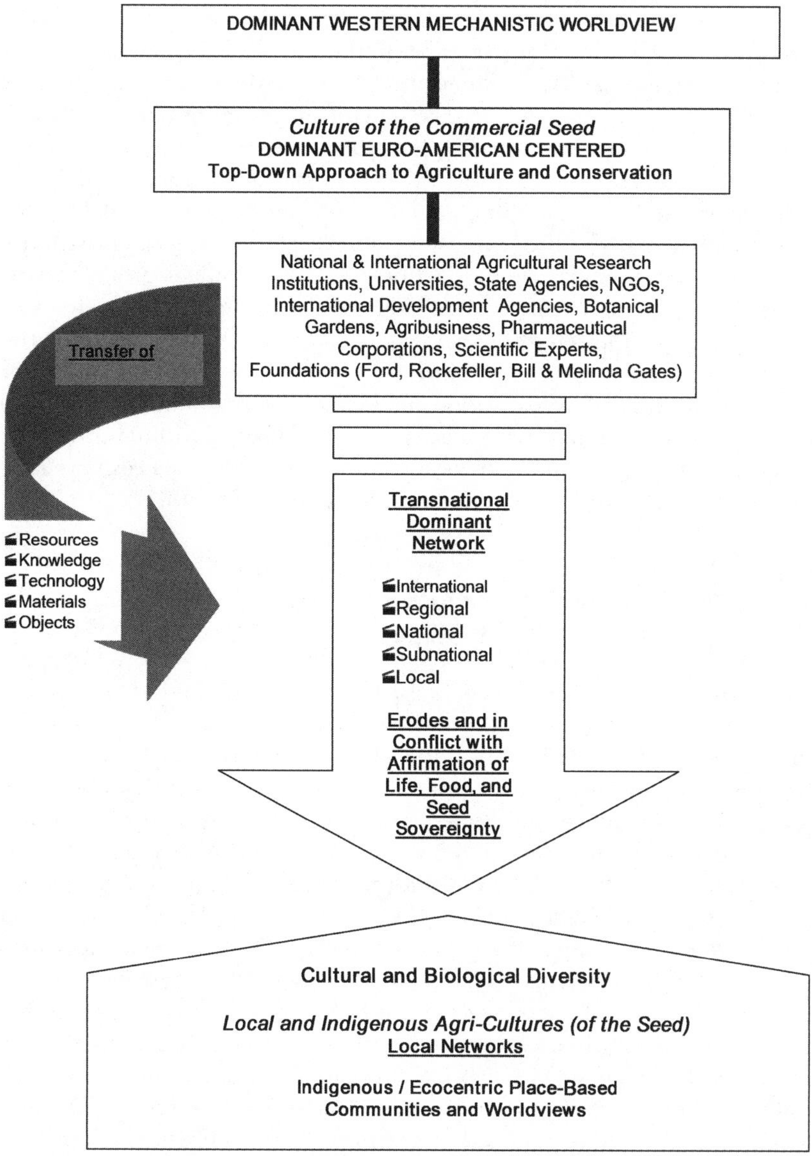

DOMINANT WESTERN MECHANISTIC WORLDVIEW

Culture of the Commercial Seed
DOMINANT EURO-AMERICAN CENTERED
Top-Down Approach to Agriculture and Conservation

National & International Agricultural Research
Institutions, Universities, State Agencies, NGOs,
International Development Agencies, Botanical
Gardens, Agribusiness, Pharmaceutical
Corporations, Scientific Experts,
Foundations (Ford, Rockefeller, Bill & Melinda Gates)

Transfer of

- Resources
- Knowledge
- Technology
- Materials
- Objects

Transnational
Dominant
Network

- International
- Regional
- National
- Subnational
- Local

Erodes and in
Conflict with
Affirmation of
Life, Food, and
Seed
Sovereignty

Cultural and Biological Diversity

Local and Indigenous Agri-Cultures (of the Seed)
Local Networks

Indigenous / Ecocentric Place-Based
Communities and Worldviews

FIGURE 4.1. Dominant conventional approach to agriculture and conservation from above. From Escobar (1998, 1999); Gonzales (1996); Gonzales, Chambi, and Machaca (1999); Gonzales et al. (2010); IAASTD (2009); and Pimbert (1994b).

This chapter explores the connections between *in situ* conservation and culture, place, land, and territory in the Americas. While I agree with Arturo Escobar's (2001) argument that "culture sits in places," I further contend that place and culture are dynamically and inextricably rooted in land and territory, spirituality, language, and worldview, and that in situ conservation is holistic and culture based. Here, I focus on the historical and contemporary implications of a global process that began with European colonization and continues today with the erasure of senses of place and a decline in cultural and biological diversity. This global process threatens and challenges local communities' efforts to recreate place from epistemologies and ontologies that differ from the dominant post-Enlightenment Eurocentric mind frame. The dominant Western strategy of conservation implies a narrow scientific approach, whereas indigenous and local conservation is embedded within fundamentally different agricultural systems, strategies, and worldviews/cosmovisions (Gonzales et al. 2010; IAASTD 2009; Sain and Calvo 2009).

Indigenous Peoples of the Americas and the Defense of Place

According to Mexican ethnoecologist Victor Toledo (2006:81–22), indigenous peoples are (1) the descendants of the original inhabitants of a territory seized by conquest; (2) "ecosystemic peoples" such as farmers, shepherds/pastoralists, hunters and gatherers, fishermen/fisherwomen, or artisans who adopt a multiple practice strategy for the use of nature; (3) collectivities that practice labor-intensive rural agriculture that produces small amounts of surplus within systems with low energy needs; (4) collectivities that do not have centralized political institutions and therefore organize their life at the community level and make decisions on the basis of consensus; (5) collectivities that share language, religion, moral values, beliefs, clothing, and other characteristics of belonging, usually in relation to a particular territory; (6) collectivities that have a different worldview, which is characterized by a non materialistic attitude; (7) collectivities that live subservient to a dominant culture and society; and (8) collectivities that are composed of individuals that subjectively consider themselves as indigenous.

In Central and South America, indigenous peoples' cultural diversity is highly correlated with biodiversity (Hawkes 1992; Fowler and Mooney 1990; Posey 1999; Toledo 2001a, 2001b, 2006; Toledo et al. 2001). Indigenous peoples live in 80 percent of protected areas in Latin America (Colchester and Gray 1998).[3] In Central and South America alone there are four hundred ethnic groups, each with its own distinct language, social organization, and cosmovision. These groups represent diverse forms of economic organization and ways of production adapted to the ecosystems that they inhabit (Deruyttere 2003; Nabhan 1985, 1997, 2002; Smith 2002; Toledo 2006). According to conservative estimates, the total indigenous population in the Americas is around 53 million people. In the United States, around 2 percent of the population (4.1 million people) is indigenous, while in Canada indigenous people make up 4.3 percent (fewer than 2 million) of the total population. The indigenous population of Mexico and Central and South America is less than 10 percent (around 48 million) of the total population (Deruyttere 2003).

The total area controlled by indigenous peoples throughout the Americas has shrunk significantly (Gilbert, Wood, and Sharp 2002; LaDuke 1990; Toledo et al. 2001) due to nation-state building and both external and internal colonialism (Gasteyer and Butler Flora 2000). We have thus seen, particularly from the 1940s onward, the outgrowth of space-oriented, nonsustainable monocultures of the mind, land, and spirit. This erosive neocolonial process has challenged sustainability rooted in indigenous places. The shrinking of native places, land, and territory in the Americas can be illustrated by the current land surface controlled by indigenous peoples in Mexico and Central America provided by Toledo and his team of researchers. These data can also be interpreted as an index of the erasure of indigenous peoples' places in this area: more than 29 million hectares (15 percent) for Mexico; more than 16 million hectares (14 percent) for Honduras; 5.9 million hectares (45.3 percent) for Nicaragua; more than 320,000 hectares (6.2 percent) for Costa Rica; and 1,657,100 hectares for Panama (22 percent) (Toledo et al. 2001:14). In reference to this issue in North America, LaDuke notes that today in the United States,

> there are over 700 Native nations, . . . [and] Native America covers 4 percent of the land, with over 500 federally recognized tribes. Over 1,200 Native American reserves dot Canada. The Inuit homeland, Nunavut, formerly one-half of the Northwest Territories, is an area of land and water, including Baffin Island, five times the size of Texas, or the size of the entire Indian Subcontinent. Eighty-five percent of the population is Native. While Native peoples have been massacred and fought, cheated, and robbed of their historical lands, today their lands are subject to some of the most invasive industrial interventions imaginable. According to the Worldwatch Institute, 317 reservations in the United States are threatened by environmental hazards, ranging from toxic wastes to clear-cuts. (1990:2)

Gilbert et al. (2002) in their study on private agricultural land by race and ethnicity note "minorities own only a small part of the U.S.

agricultural land base. American Indians own 0.4 percent of the total of this land base" (56).[4]

According to Western standards, indigenous peoples are among the poorest of the poor (Deruyttere 1997; Hall and Patrinos 2005; Psacharopoulos and Patrinos 1994; Smith 2002), and most are socially, politically, economically, and ethnically marginalized by the nation-state and dominant society (Smith 2002; Stavenhagen 1990, 2002). Despite these commonalities, a single definition of indigenous or native peoples has not emerged within or outside the indigenous world. Rather, many social categories and critical metaphors have shaped current thinking and emphasized the complex interaction of many social actors. In light of hegemonic ideologies that devalue indigeneity,[5] or colonization of the mind, some indigenous people prefer to be called *campesinos* or peasants.[6] Others assertively call themselves *pueblos originarios*, or original peoples. International organizations like the United Nations, the World Bank, the Interamerican Development Bank, and the Organization of American States are increasingly using the term indigenous peoples as it has been proposed in the 2007 U.N. Declaration on the Rights of Indigenous Peoples and International Labor Organization (ILO) Convention 169, also known as the Indigenous and Tribal Peoples Convention. However, international law scholar James Anaya (2000) points out that despite all the benefits related to indigenous rights in the Indigenous and Tribal Peoples Convention, the implication of the term "peoples" has been curtailed by state members within the ILO Convention 169 due to fear that "it may imply an effective right of secession." Also, for that reason, "the term *self-determination* in this connection continued to raise controversy" (Anaya 2000:49; Smith 2002). However, even in the most radical Latin American country cases (Nicaragua, Bolivia, and Ecuador), there is no evidence that indigenous peoples are aiming for withdrawal from their respective nation-states (Gonzalez, Burguete Cal y Mayor, and Ortiz-T. 2010).

Critical theories such as political economy and political ecology are still weak on the ethnic/indigenous and nature question and the terminology associated with it (O'Connor 1991, 1992, 1993; Gonzales 1996; Stavenhagen 1990).[7] The term "indigenous peoples," like the terms "peasant," "campesino," "native," "aboriginal," "ladino," "mestizo," "mixed blood," "misti," "cholo," and "indio," has been used by colonizers and their descendants to divide and dominate the colonized (Forbes 1983).[8] Uncritical acceptance of such terms favors the perpetuation of colonization and the alienation of the individuals and communities marginalized by these terms. Nevertheless, in many colonized regions indigenous peoples have appropriated the term that portends to define them. To some extent, they resist colonization by affirming their cultures and places, often exerting conscious effort toward decolonizing/reaffirming their cultures and identities and giving a new twist to the colonizer's racist/ assimilationist terminology. In other cases, some colonized indigenous peoples may assume certain traits that characterize the colonizer, creating a hybrid, at times conflicted identity. The term "indigenous" can also refer to those individuals and social collectivities that have refused to play the role of the colonized or the colonizer by means of cultural affirmation.[9] Detachment from the colonizer/colonized dichotomy may occur given that both colonizer and colonized have been alienated from their real identity and an ecospiritual centered worldview.

The majority of Westerners and western institutions (civil society, politicians, public servants, nongovernmental organizations, scientists) are not aware of the significance and value of indigenous peoples' cultural diversity, and worldviews/cosmic visions, their long-term coevolution with biological diversity, and what that diversity may contribute to achieving internal balance and procuring balance of life as a whole (Nakashima et al. 2012; Salick and Liverman 2007). As a matter of fact, their disappearance will be an enormous loss with serious implications on key fronts—dynamic and innovative place-based, culturally sensitive, and resilient strategies and technologies to deal with biodiversity conservation, ecosystem management, food security, and climate change. Despite the fact that today, in different degrees, the majority of indigenous communities are in a fragile state due to the devastating consequences of five hundred years of erosive colonial and post/neocolonial regimes and policies, indigenous/peasant cultures hold at least part of the answer to critical issues not yet resolved by Western societies (Nakashima et al. 2012; Salick and Liverman 2007). Recently the ETC Group (2009) has noted that indigenous peasants feed at least 70 percent of the world's population. More than 20 million indigenous peasants produce 60 percent of Latin America's total food output (Altieri and Koohafkan 2008), and Peruvian Andean indigenous farmers produce 60 percent of Peru's total food output (Maffi and Woodley 2010). These issues of inequality have a lot of implications for nationhood, reciprocity, justice, spirituality, illness and healing, environmental and ecological balance, culture, morals, ethics, economy, and sustainability (Mander and Tauli-Corpuz 2006; Merchant 1992; Tauli-Corpuz et al. 2010).

Divergent Worldviews

In contrast to the dominant worldview and its mechanistic paradigm of development, the indigenous worldview is not based on a dichotomizing principle. It promotes an intimate relationship to land and territory, that is, to place, locality, and habitat (Gray 1999; Maffi 2001; Nabhan 1997, 2002; Toledo et al. 2001). Indigenous peoples tend to see themselves not outside of nature but, rather, as part of it and as procuring balance and harmony from within it (Gonzales and Gonzalez 2010; Gonzales and Nelson 2001; LaDuke 1994; Pilgrim and Pretty 2010; Posey and Dutfield 1996). Spirituality and ritual practices that sustain the environment are also part of their integral and complex cosmologies (Gonzales and Gonzalez 2009, 2010; Gray 1999; Posey 1999). Rafael Guitarra, an Ecuadorian Quichua, alludes to this when he stresses the importance of the relationship between human beings and nature within a human cosmovision that is integral, guiding all decisions and strategies: "Because of this, we are not able to speak of 'conservation' alone but always of coexistence. This is how we live in the indigenous world, together with and as part of nature, sharing the sentiment of Mother Earth (*Pachamama*). If we do not take care of and nurture Mother Earth, what can we expect of her?" (Nazarea and Guitarra 2004:18).

Place embodies the intimate experiential interaction of a community of human beings with each other and with their local environment. According to Native American scholar Gregory Cajete, the concept of place is often taken for granted. In contemporary Western societies the notion of place is a given; when most Western people speak about a place, they assume that everyone has the same reference to that place. Thus, "in the Western scientific perspective, maps of places are drawn to symbolically represent a place based on previously agreed upon criteria that are logical and measurable with regard to the discipline of cartography. But a map is always just a kind of symbol for a place; it is not the place it is meant to describe. Indeed to know any kind of physical landscape you have to experience it directly; that is, to truly know any place you have to live in it and be part of its life process" (Cajete 1999:181). Place is where identity and language are forged through interaction with the natural and supernatural, in local stories and histories that are interwoven and recreated across the landscape (Cajete 2000, 2001; LaDuke 1994; Merchant 1992).

Escobar argues that place is culturally and historically specific and points out that while "place has dropped out of sight in the 'globalization craze' of recent years, it has become evident that this erasure of place has profound consequences for our understanding of culture, knowledge, nature, and economy" (Escobar 2001:141). Place does not mean the same to an Andean Quechua or an Aymara as it does to a conventional farmer from Kentucky, Iowa, or Kansas, or, for that matter, to an agroecologist, a conservationist, or organic market niche-oriented farmer from California (Gonzales et al. 2010). Historical specificity shapes each experience of place. For indigenous peoples, "the geography, the local and the communal, the neighborhood and the house, have doubtless a very different

Table 4.1. Two contemporary templates of conservation

Western Mechanistic Worldview: (Neo)Colonizer's Model (from Above)	Indigenous Place-Based Model (from Below)
1. Western epistemology, ontology, cosmovision.	1. Indigenous peoples' epistemologies, ontologies, cosmovisions.
2. Grounded in the Judeo-Christian and Cartesian cosmovision.	2. Grounded in indigenous, precolonial cosmovision.
3. Man dissociates from nature (subject-object).	3. Human beings are part of life as a whole (we all are but one).
4. Anthropocentric vision of the world: man is the center of the world.	4. Human beings are part of a community of equivalents.
5. Mechanistic worldview.	5–9. Multiple interaction among the community of human beings, the community of nature, and the community of the deities/gods. Their relation is among equivalents. All beings are incomplete, therefore the possibility of complementing each other and sharing. Verb based. Knowledge/saberes are held temporarily and circulate through the community of human beings. In this view everything is alive.
6. Life moves around men's material needs.	
7. Egocentric ethic: what is best for the individual is best for society as a whole.	
8. Based on Western mechanistic science and capitalism; lab and noun based.	
9. Earth is dead and inert, manipulable from outside, and exploitable for profits.	
10. Innovation protected by individual property rights.	10. Innovation takes place within the interaction of the three major communities, emerges within a tradition.
11. Linear vision of history (past-present-future).	11. Circular vision of history.
12. Specialized/fragmented.	12. Holistic.
13. Space; homogenizing/standardizing.	13. Place-diversity oriented.
14. Nonsustainable.	14. Sustainable.
15. Sustainability concept is foreign to this current dominant worldview and its development paradigms.	15. Sustainability is incorporated in the worldview. Throughout the year, rituals and ceremonies contribute to procure it.

Adapted from Merchant (1992), Gonzales and Gonzalez (2010), and IAASTD (2009).

gravitational pull than in the case of disperse, sometimes itinerant or . . . urban populations from industrial societies" (Quijano 2004:14). For Quechuas, Kichwas, and Aymaras, for instance, place is the local *Pacha* (mother earth), where the three communities of nature (*Sallqa*), human beings (*runas*), and deities interact (Gonzales, Chambi, and Machaca 1999; Gonzales and Gonzalez 2010; IAASTD 2009) (see table 4.1).

Place, like culture and language, is intertwined with land and the cosmos. For Andean indigenous people, Pacha is the local representation of the macrocosm at the microlevel (Gonzales and Gonzalez 2010). The construction of place and culture is not possible without a strong connection to land. Through time, however, peoples' connection to place may become unstable and disconnected from its foundation on the land and devoid of its multiple connections and interactions with habitat. The process of cultural "hybridization" may imply, over time, a temporary or permanent disconnect from a nurturing and sustainable life. The contemporary Western worldview regards land in a fragmented and reductionist way, detached from its multiple connections and devoid of holistic meaning. Today, globalization is aggressively threatening what remains of place and place making on Earth. For small-scale farmers in rural areas of North America, for indigenous peoples and First Nations everywhere, and for those who are struggling to "become native to [a] place" and to this land (Jackson 1996), land remains the fundamental element of the reproduction of their subaltern cultures. Because unreconciled land issues (Van Dam 1999) drive the "indigenous problem" (Quijano 2004) and the "ethnic question" (Stavenhagen 1990), the question of place and culture in North America and elsewhere has not been substantially resolved.

The Cultures of the Seed

In line with the divergence between Western and indigenous worldviews, two principle viewpoints influence the relations of agrobiodiversity: (1) the "culture of the commercial seed" embedded within contemporary Western dominant institutions, networks, discourses, social relations, and agriculture; and (2) the "culture of the native seed," which nests within the wide and rich diversity of indigenous peoples and peasant cultures (Gonzales, Chambi, and Machaca 1999). The concept of "cultures of the seed" helps elucidate the connection between the persistence or erasure of place and the conservation of agrobiodiversity. The divergence between the "two cultures" highlights the fact that the seed is not simply a raw material that plant breeders can manipulate in a laboratory or commodity we buy at a seed store. The cultures of the seed reference specific cosmological views and cognitive models, diverse technological strategies and ecosystems, agricultural systems, and substantially different types of social, religious, and productive organizations and rituals (Asociación Bartolomé Aripaylla 2001; Chambi and Chambi 1995; Gonzales 1996; IAASTD 2009; Nabhan 1997; Sain and Calvo 2009; Toledo et al. 2001).

The culture of the native seed *preconfigures* the commercial seed. It hosts open-pollinated varieties ("landraces" in scientific terms). Landraces are commonly known as traditional varieties of plants or animals that have developed without excessive modification, adapting more naturally to the local environment. Yet from the international level to the local level, commercial seed (including hybrid seeds and genetically modified seeds) restricts and threatens the cultures of the native/local seed. Hybrid seed results from cross-pollinating compatible varieties of plants. Vicki Mattern (2013) notes that, "unlike hybrids which are developed in the field using natural, low-tech methods, GM varieties are created in a lab using highly complex technology, such as gene splicing. These high-tech GM varieties can include genes from several species—a phenomenon that almost never occurs in nature."

Mostly under the control of agribusiness and corporations, hybrid seeds represent the outcome of scientific manipulation that narrowly privileges certain genetic traits. R. P. Haynes notes that the commercial seed has "tended to be monocultural both in the sense that it is intolerant of alternative cultures and in the sense that it relies heavily on monocultural cropping systems" (1985:1). Plant breeding is carried out in an isolated space, removed from the agricultural place, the local community, and the ecological niche. The culture of the commercial seed has given birth first to a modern technological revolution in modern agriculture, which from the 1960s onward has been exported to Global South countries as the "Green Revolution." The second "revolution" is now associated with the spread of biotechnology and genetically modified organisms (GMOs), including transgenic crops.

The consequences of transgenics for the entire ecological chain will not be seen immediately, but they may contribute to health risks and environmental pollution for the planet as a whole. Consequently, the spread of the second "revolution" is also promoting the growth of space, the erosion of place and non-Western indigenous cultures, and ecocide. Embedded within non-Western cosmologies, the native seed is the outcome of a nurturing process, rooted in an intimate mental, physical, social, and spiritual relationship between humans, their crops, and local environments. This nurturance takes place on plots of land—the *milpas* of Mesoamerica, the *chacras* of the Andes-Amazonian region, and the *finca* in Guna Yala, Panama. In milpas, fincas, and chacras a relational cultural complex forms through ritual with the spiritual community of the deities and gods, the community of human beings, and the community of nature. In this context, all entities are living beings that treat each other with respect, coequal and interactive. In the agricultural cycle, rituals are the passages through which harmony, balance, and life are procured. Guatemalan activist and Nobel Peace Prize laureate Rigoberta Menchu highlights the intimate connections between indigenous peoples and the native seed within the context of indigenous baptism: "The child is present for all of this, although he's all wrapped up and can scarcely be seen. He is told that he will eat maize and that, naturally, he is already made of maize because his mother ate it while he was forming in her stomach. He must respect the maize; even the grain of maize which has been thrown away, he must pick up" (1992:13).

In sum, the culture of the commercial seed is part and parcel of the mechanistic and fragmented worldview, while the culture of the native seed embraces an indigenous holistic worldview (see table 4.1). On one hand, the culture of the commercial seed fosters a transnational network (see figure 4.1) "that encompasses diverse sites in terms of actors, discourses, practices, cultures, and stakes" (Escobar 1999:1). This transnational network has encouraged the outgrowth of monocropping, monocultures of the mind and the land, and, as a consequence, the dominance of space over place. On the other, the culture of the native seed fosters above all a sense of place and a nurturing network that encompasses diverse places (see chapter 12, this volume). Its central tenet is the procurement of harmony and balance with life as a whole. An almost natural outcome is the increase of diversity as a whole.

Conclusion

The strongholds of place and *in situ* conservation in the Americas are indigenous peoples (peasant farmers in particular), as well as rural and urban farming communities and their unique worldviews. They constitute an alternative *in situ* conservation strategy based on their traditional seed systems. Indigenous peasant *in situ* conservation is a window of opportunity for working toward cultural affirmation and the regeneration of the biocultural-spiritual landscape as a whole (Ishizawa 2009; Ishizawa et al. 2010). The problem is that these communities are not only marginalized, but many of them are tackling nearly overpowering political and socioeconomic crises. Indigenous autonomy and self-determination, control over natural resources, and indigenous land and territory are central issues that are still not satisfactorily resolved. It is of the utmost importance to make the U.N. Declaration on the Rights of Indigenous

Peoples and other relevant covenants (e.g., U.N. Convention on Biological Diversity, ILO 169) work. Solid efforts at place/culture/cultural affirmation/reindigenization/community making, despite their vitality, innovativeness, and solid proposals (e.g., PRATEC-NACA in Peru, AGRUCO in Bolivia), are still small in scale due to a lack of government and international cooperation and support through culturally sensitive policies and projects. New culture-sensitive paradigms/approaches and stronger alliances, at the hemispheric, regional, national, and local levels, can and should be built among those concerned with community resilience and the erasure of place and what this represents for human cultural and biological diversity.

There is enough land in the Americas for the affirmation of place from subaltern discourses, practices, paradigms, and worldviews that challenge hegemonic neocolonial forces. There is enough land to produce enough food for and by the hungry under a more holistic and ecocentric paradigm. But in many regions of the Americas, there is a pressing need for culturally appropriate, participatory agrarian reforms consistent with indigenous and ecocentric values. This process should start with the individual and with communities at the local level. Community-based decolonization, the deconstruction of modernization and development, is the passage toward place-based cultural affirmation. For some individuals and communities, this process will be a strengthening of their places; for others it will be a process of moving out of their "violent environments" (Peluso and Watts 2001) and becoming native to their places (Jackson 1996). Land, territory, spirituality, and institutions are central to this process. Decolonization embraces both indigenous and nonindigenous peoples and is proposed as an active response to the ongoing erasure of place by the forces representing globalization. Of primary importance to reindigenization—becoming native to a place—is the restoration of plant and animal diversity characteristic of these places and cultures and the world's diverse ecosystems. This would definitely be a step toward a more balanced reconnection with Mother Earth as a whole.

Notes

1. A general definition of *worldview* is the basic ways of seeing, feeling, and perceiving the world. It manifests in the ways in which a people act and express themselves.

2. Battiste and Henderson note that Eurocentrism "is the imaginative and institutional context that informs contemporary scholarship, opinion, and law. As a theory, it postulates the superiority of Europeans over non-Europeans. It is built on a set of assumptions and beliefs that educated and usually unprejudiced

Europeans and North Americans habitually accept as true, as supported by the 'facts,' or as reality" (2000:21).

3. In Latin America, protected areas and indigenous peoples are the current conceptualization of what has resulted from the effects of colonial and postcolonial patriarchal, mainstream development-oriented, anti-ethnic, anti-environment, racist regimes. While indigenous peoples are the remains of a violent process of exploitation, marginalization, and dislocation, to which local indigenous populations and their territories have been submitted, protected areas are the remains of last-century state colonization and development policies. Protected areas are part of a wider set of policies and activities, more or less defined and controlled by the state, related to land use and management of natural resources (Colchester and Gray 1998; Diegues 2000).

4. On the issue of erasure of place in North America, see Churchill (1993) and Gilbert et al. (2002).

5. For a reflection on this topic, see Armstrong (2010).

6. The campesino/peasant concept originates in Europe and stresses class over ethnicity. It conflicts with indigenous people(s)' concepts. Development and Marxist theories neglect ethnicity/identity/culture. Official national and international statistics (Food and Agriculture Organization of the United Nations, Interamerican Development Bank, World Bank) illustrate this. The concept favored the old, unproductive "peasantization-depeasantization" debate. For Grimaldo Rengifo, "although the word *campesino* literally translates to «peasant» the connotation of the word is somewhat different. The *campesino* is a social phenomenon specific to Latin America. Thus, the word is preserved in its original form of *campesino* throughout the text" (2005:1).

7. From an indigenous (studies) perspective the theoretical and epistemological weakness on the ethnic/indigenous and nature question has been debated without much success in the journal *Capitalism, Nature and Socialism* (O'Connor 1991, 1992, 1993).

8. For a discussion in English on the mestizo concept see Forbes (2013), http://inxinachtliinmilpa.blogspot.ca/2013/01/the-mestizo-concept-product-of -european.html, accessed May 5, 2013.

9. Cultural affirmation for the Andes is illustrated by PRATEC-NACA's movement. This cluster of Andean indigenous NGOs is one concerned with the affirmation of life as a whole, in order to nurture a culturally diverse world through the recovery and revitalization of agriculture and cattle raising and cultural practices of Andean-Amazonian indigenous peoples. It is aligned with the interests of indigenous communities in a process of *acompañamiento*, walking side by side with, and facilitating the collection and systematization of indigenous epistemologies (Ishizawa 2009; Rengifo 2001; Valladolid and Apffel-Marglin 2001b).

References

Altieri, M., and P. Koohafkan. 2008. *Enduring Farms: Climate Change, Smallholders and Traditional Farming Communities*. Environment and Development Series 6. Geneva: Third World Network.

Anaya, S. James. 2000. *Indigenous Peoples in International Law*. New York: Oxford University Press.

Armstrong, J. 2010. Indigeneity: The heart of development with culture and identity. In *Towards an Alternative Development Paradigm: Indigenous Peoples Self-Determined Development*, edited by Tebtebba Foundation, 79–88. Baguio City, Philippines: Indigenous Peoples' International Centre for Policy Research and Education.

Asociación Bartolomé Aripaylla. 2001. *Kawsay, Kawsaymama: La regeneración de semillas en los andes centrales del Peru. El caso de la Comunidad quechua de Quispillacta*. Ayacucho: Asociación Bartolomé Aripaylla.

Battiste, Marie, and James [Sa'ke'j] Youngblood Henderson. 2000. Eurocentrism and the European ethnographic tradition. In *Protecting Indigenous Knowledge and Heritage: A Global Challenge*, 21–56. Saskatoon: Purich.

Cajete, Gregory. 1999. A sense of place. In *Native Science: Natural Laws of Interdependence*, 177–213. Santa Fe, NM: Clear Light.

———. 2000. Indigenous knowledge: The Pueblo metaphor of indigenous education. In *Reclaiming Indigenous Voice and Vision*, edited by Marie Battiste, 192–208. Toronto: UBC Press.

———. 2001. Indigenous education and ecology: Perspectives of an American Indian educator. In *Indigenous Traditions and Ecology: The Interbeing of Cosmology and Community*, edited by John Grim, 619–638. Cambridge, MA: Harvard University Press.

Chambi, Nestor, and Walter Chambi. 1995. *Ayllu y Papas: Cosmovision, religiosidad y agricultura en Conima*. Puno, Peru: Asociación Chuyma de Apoyo Rural "Chuyma Aru."

Churchill, Ward. 1993. *Struggle for the Land: Indigenous Resistance to Genocide, Ecocide, and Colonization*. Monroe, ME: Common Courage Press.

Colchester, M., and A. Gray. 1998. Foreword. In *From Principles to Practice: Indigenous Peoples and Biodiversity Conservation in Latin America*, edited by Andrew Gray, Alejandro Parellada, and Hellen Newing, 17–20. Proceedings of the Pucallpa Conference, Pucallpa, Peru, March, 1997. IWGIA Document 87. Copenhagen: International Work Group for Indigenous Affairs.

Declaration of Atitlán. 2002. Indigenous Peoples' Consultation on the Right to Food: A Global Consultation. Atitlán, Panajachel, Sololá, Guatemala, April 17–19.

Declaration of Nyéléni. 2007. Declaration of the Forum for Food Sovereignty. Nyéléni Village, Sélingué, Mali.

Deruyttere, Anne. 1997. *Indigenous Peoples and Sustainable Development: The Role of the Inter-American Development Bank.* Inter-American Development Bank Publication 16. Washington, DC: Inter-American Development Bank.

———. 2003. *Pueblos Indígenas, recursos naturales y desarrollo con identidad: Riesgos y oportunidades en tiempos de globalización.* Unidad de Pueblos Indígenas y Desarrollo Comunitario. Departamento de Desarrollo Sostenible. Washington, DC: Inter-American Development Bank.

Diegues, Antonio C. 2000. *El Mito Moderno de la Naturaleza Intocada.* Quito, Ecuador: Ediciones Abya Yala.

Escobar, A. 1995. *Encountering Development. The Making and Unmaking of the Third World.* Princeton, NJ: Princeton University Press.

———. 1998. Whose knowledge, whose nature?: Biodiversity, conservation, and the political ecology of social movements. *Journal of Political Ecology* 5:53–82.

———. 1999. Biodiversity: A perspective from within. *Seedling* 3.

———. 2001. Culture sits in places: Reflections on globalism and subaltern strategies of localization. *Political Geography* 20:139–174.

ETC Group. 2009. Who will feed us? Questions for the food and climate crises. *ETC Group Communiqué,* November, no. 102.

Forbes, Jack. 1983. El concepto de mestizo-metis. *Plural Segunda Epoca* Vol. 13–14, no. 145, October: 20–29.

———. 2013. The Mestizo concept: A product of European imperialism. http:// inxinachtliinmilpa.blogspot.ca/2013/01/the-mestizo-concept-product-of-euro pean.html, accessed May 5, 2013.

Fowler, Cary, and Pat Mooney. 1990. *Shattering: Food, Politics, and the Loss of Genetic Diversity.* Tucson: University of Arizona Press.

Garavito, Cesar. 2011. *El Derecho en America Latina: Un mapa para el pensamiento jurídico del siglo XXI.* Buenos Aires: Siglo Veintiuno Editores.

Gasteyer, Stephen P., and Cornelia Butler Flora. 2000. Modernizing the savage: Colonization and perceptions of landscape and lifescape. *Sociologia Ruralis* 40(1):128–149.

Gilbert, Jess, Spencer Wood, and Gwen Sharp. 2002. Who owns the land?: Agricultural land ownership by race and ethnicity. *Rural America* 17(4):55–62.

Gonzalez, M., A. Burguete Cal y Mayor, and P. Ortiz-T. 2010. *La autonomía a debate: autogobierno indígena y Estado plurinacional en América Latina.* Sede, Ecuador: FLACSO (Facultad Latinoamericana de Ciencias Sociales).

Gonzales, T. 1996. Political ecology of peasantry, the seed, and NGOs in Latin America: A study of Mexico and Peru, 1940–1996. Ph.D. dissertation, University of Wisconsin, Madison.

———. 2000. The cultures of the seed in the Peruvian Andes. In *Genes in the Field: On-Farm Conservation of Crop Diversity,* edited by Stephen B. Brush.

Rome, ON: International Development Research Centre (Canada); International Plant Genetic Resources Institute.

Gonzales, T., N. Chambi, and M. Machaca. 1999. Agricultures and cosmovision in contemporary Andes. In *Cultural and Spiritual Values of Biodiversity*. London: Intermediate Technology.

Gonzales, T., and M. Gonzalez. 2009. The spirit of sustainability and the *Ayllu* in South America. In *The Spirit of Sustainability: Forum on Religion and Ecology*, edited by Willis Jenkins, 233–235. New Haven, CT: Yale University Sustainability Project.

———. 2010. From colonial encounter to decolonizing encounters. Culture and nature seen from the *Andean Cosmovision of Ever*: The nurturance of life as whole. In *Nature and Culture*, edited by J. Pretty and S. Pilgrim, 83–101. London: Earthscan.

Gonzales, Tirso, Marcela Machaca, Nestor Chambi, and Zenon Gomel. 2010. Latin American Andean indigenous agriculturalists challenge the current transnational system of science, knowledge and technology for agriculture: From exclusion to inclusion. In *Towards an Alternative Development Paradigm. Indigenous Peoples Self-determined Development*, edited by Victoria Tauli-Corpuz, Leah Enkiwe-Abayao and Raymond de Chavez, 163–204. Baguio City, Philippines: Tebtebba Foundation.

Gonzales, Tirso, and Melissa Nelson. 2001. Contemporary Native American responses to environmental threats in Indian Country. In *Indigenous Traditions and Ecology: The Interbeing of Cosmology and Community*, edited by John A. Grim, 495–538. Cambridge, MA: Harvard University Press.

Gray, Andrew. 1999. Indigenous peoples, their environments and territories. In *Cultural and Spiritual Values of Biodiversity*, edited by Darrell Posey, 61–66. U.N. Environmental Programme on Global Biodiversity Assessment. Cambridge, MA: Cambridge University Press.

Hall, G., and H. Patrinos. 2005. *Pueblos indígenas, pobreza y desarrollo humano en América Latina: 1994–2004*. Washington, DC: Banco Mundial.

Hawkes, J. 1992. The centers of plant genetic diversity in Latin America. *Diversity*: 7(1 and 2):7–9.

Haynes, R. P. 1985. Agriculture in the US: Its impact on ethnic and minority groups. *Agriculture and Human Values* 3:1–3.

IAASTD (International Assessment of Agricultural Knowledge, Science and Technology for Development). 2009. *Agriculture at a Crossroads*, Vol. 3: *International Assessment of Agricultural Science and Technology for Development: Latin America and the Caribbean*. Washington, DC: Island Press.

Indigenous Environmental Network. 2001. Indigenous environmental network statement on the right to food and food security. Paper presented at Indigenous Environmental Network's 12th annual Protecting Mother Earth Conference, What We Do Now Touches the Next Seven Generations. Penticton Indian

Band Okanagan Nation Territories, Penticton, British Columbia, Canada, August 2–5, 2001.

IPAF (Indigenous Partnership for Agrobiodiversity and Food Sovereignty). 2010. Scoping report, with the support of the Christensen Fund, and in collaboration with the International Institute for Environment, Slow Food International and three indigenous organizations (Tebtebba Foundation, ANDES, and the Vanuatu Cultural Centre). Pisaq, Cusco, Peru, May 3–5, 2010.

Ishizawa O., J. 2003. Criar diversidad en los Andes del Perú. Los desafíos globales. Kawsay mama (3). PRATEC (Proyecto Andino para las Tecnologías Campesinas [Andean Project for Peasant Technologies]). Peru: PRATEC.

———. 2009. Affirmation of cultural diversity—learning with the communities in the central Andes. Draft Thematic Paper, What Next Forum. *Development Dialogue 2.*

Ishizawa O., J., G. Rengifo, and N. Arnillas. 2010. La Crianza del Clima en los Andes Centrales del Peru. El papel del Fondos de Iniciativas de Afirmacion Cultural en la regeneracion del Allin Kawsay o vida buena andina. Peru: PRATEC Proyecto Andino para las Tecnologías Campesinas (Andean Project for Peasant Technologies).

Iturralde, D. and E. Krotz. 1996. *Indigenous Development: Poverty, Democracy and Sustainability.* December, No. IND96-102. Washington, DC: InterAmerican Development Bank.

Jackson, Wes. 1996. *Becoming Native to this Place.* Washington, DC: Counterpoint.

LaDuke, Winona. 1990. *All Our Relations: Native Struggles for Life and Land.* Cambridge, MA: South End Press.

———. 1994. Traditional ecological knowledge and environmental futures. *Colorado Journal of International Environmental Law and Politics* 5:127–148.

Lander, Edgardo. 2000. Eurocentrism and colonialism in Latin American social thought. *Nepantla: Views from South* 1(3):519–532.

Maffi, L. 2001. *On Biocultural Diversity: Linking Language, Knowledge, and the Environment.* Washington, DC: Smithsonian Institution Press.

Maffi, L., and E. Woodley. 2010. *Biocultural Diversity Conservation. A Global Sourcebook.* London: Earthscan.

Mander, J., and Tauli-Corpuz, V. 2006. *Paradigm Wars: Indigenous Peoples' Resistance to Globalization.* San Francisco: Sierra Club Books.

Mattern, V. 2013. Hybrid Seeds vs GMOs. January 20, http://www.motherearth news.com/real-food/hybrid-seeds-vs-gmos-azb0z1301zsor.aspx#axzz2SjF6Dn Da, accessed May 5, 2013.

Menchu, Rigoberta. 1992. *I, Rigoberta Menchu: An Indian Woman in Guatemala.* Edited by E. Burgos-Debray. New York: Verso.

Merchant, Carolyn. 1992. *Radical Ecology: The Search for a Livable World.* New York: Routledge.

Nabhan, Gary. 1985. Native American crop diversity, genetic resource conservation, and the policy of neglect. *Agriculture and Human Values* 3:14–17.

———. 1997. *Cultures of Habitat: On Nature, Culture, and History.* Washington, DC: Counterpoint.

———. 2002. *Enduring Seeds: Native American Agriculture and Wild Plant Conservation.* Tucson: University of Arizona Press.

Nakashima, D. J., K. Galloway McLean, H. D. Thulstrup, A. Ramos Castillo, and J. T. Rubis. 2012. *Weathering Uncertainty: Traditional Knowledge for Climate Change Assessment and Adaptation.* Paris/Darwin: UNESCO/United Nations University.

Nazarea, Virginia D., and Rafael Guitarra. 2004. *Stories of Creation and Resistance.* Quito, Ecuador: Ediciones Abya Yala.

O'Connor, James. 1991. Theoretical notes. On the two contradictions of capitalism. *Capitalism Nature Socialism* 2(8):107–109.

———. 1992. Symposium. The second contradiction of capitalism. *Capitalism Nature Socialism* 3(3):77–99.

———. 1993. Symposium. The second contradiction of capitalism. *Capitalism Nature Socialism* 4(1):99–108.

Pajares, Erik. 2004. Políticas y legislación en agrobiodiversidad. Kawsay mama (6). PRATEC (Proyecto Andino para las Tecnologías Campesinas [Andean Project for Peasant Technologies]). Lima, Peru: PRATEC.

Peluso, Nancy L., and Michael J. Watts. 2001. *Violent Environments.* Ithaca, NY: Cornell University Press

Pilgrim, S., and J. Pretty, eds. 2010. *Nature and Culture.* London: Earthscan.

Pimbert, Michel. 1994a. Editorial. *Etnoecologica* 2(3):3–5.

———. 1994b. The need for another research paradigm. *Seedling* 11:20–32.

———. 2006. *Transforming Knowledge and Ways of Knowing for Food Sovereignty.* London: International Institute for Environment and Development.

Posey, Darrell A. 1999. Introduction: Culture and nature—the inextricable link. In *Cultural and Spiritual Values of Biodiversity. UNEP's Global Biodiversity Assessment Volume,* edited by Darrell Posey, 3–18. Cambridge, MA: Cambridge University Press.

Posey, Darrell, and Graham Dutfield. 1996. *Beyond Intellectual Property, Toward Traditional Resource Rights for Indigenous Peoples and Local Communities.* Ottawa: International Development Research Centre.

PRATEC (Proyecto Andino para las Tecnologías Campesinas [Andean Project for Peasant Technologies]). 2006a. Núcleos de afirmación cultural Andina (ABA, APU, AWAY, NUVICHA, ARAA/CHOBA CHOBA Y PRADERA): Reflexiones sobre el Proyecto Conservación In Situ de los Cultivos nativos y sus parientes silvestres (2001–2005). Kawsay mama (10), Tomo I. Lima, Peru: PRATEC.

————. 2006b. Núcleos de afirmación cultural Andina (Chuyma Aru, Paqalqu, Qolla Aymara y Asap): Reflexiones sobre el Proyecto Conservación In Situ de los Cultivos nativos y sus parientes silvestres (2001–2005). Kawsay Mama (11), Tomo II. Peru: PRATEC.

Psacharopoulos, G., and H. Patrinos, eds. 1994. *Indigenous People and Poverty in Latin America: An Empirical Analysis*. Washington, DC: World Bank.

Quijano, Anibal. 2000. Coloniality of power, Eurocentrism and Latin America. *Nepantla: Views from the South* 1(3):533–580.

————. 2004. El "Movimiento Indigena" y las cuestiones pendientes en America Latina. Paper presented at Latin@s in the World System: Political Economy of the World Systems XXVIII Annual Conference (Section of the American Sociological Association), April 23–24, Department of Ethnic Studies, University of California, Berkeley.

Rengifo, Grimaldo. 2001. Saber local y la conservación de la agrobiodiversidad Andino-Amazónica. Kawsay mama (2). PRATEC (Proyecto Andino para las Tecnologías Campesinas [Andean Project for Peasant Technologies]). Lima, Peru: PRATEC.

————. 2003. Agrobiodiversidad y cosmovisión Andina. Kawsay mama (4). PRATEC (Proyecto Andino para las Tecnologías Campesinas [Andean Project for Peasant Technologies]). Lima, Peru: PRATEC.

————. 2004. Saber local y conservación in situ de plantas cultivadas y sus parientes silvestres. Kawsay mama (7). PRATEC (Proyecto Andino para las Tecnologías Campesinas [Andean Project for Peasant Technologies]). Lima, Peru: PRATEC.

————. 2005. The educational culture of the Andean-Amazonian community. *INTERculture* 148:1–36.

————. 2010. *Crisis Climática y Saber Comunero en los Andes del Sur Peruano*. PRATEC (Proyecto Andino para las Tecnologías Campesinas [Andean Project for Peasant Technologies]). Lima, Peru: PRATEC.

Sain, G., and G. Calvo. 2009. *Agriculturas de América Latina y el Caribe*. Elementos para una contribución al desarrollo sostenible. San José, Costa Rica: IICA (Instituto Interamericano de Cooperacion para la Agricultura), UNESCO.

Salick, Jan, and Diana Liverman. 2007. Indigenous peoples and climate change. Paper presented at Indigenous Peoples and Climate Change, April 12–13, 2007, Environmental Change Institute, University of Oxford.

Santa Fe Board of County Commissioners. 2007. Santa Fe County approves seed sovereignty resolution. No GE Seeds! May 30, http://portland.indymedia .org/en/2007/02/354468.shtml, accessed May 8, 2013.

Smith, Richard. 2002. *A Tapestry Woven from the Vicissitudes of History, Place and Daily Life: Envisioning the Challenges for Indigenous Peoples of Latin America in the New Millennium*. Lima, Peru: Ford Foundation and Oxfam America.

Stavenhagen, Rodolfo. 1990. *The Ethnic Question.* United Nations University Press.

———. 2002. The return of the native: The indigenous challenge in Latin America. Occasional paper. University of London, Institute of Latin American Studies.

Tauli-Corpuz, V., L. Enkiwe-Abayao, and R. de Chavez, eds. 2010. *Towards an Alternative Development Paradigm. Indigenous Peoples' Self-Determined Development.* El Baguio, Philippines: Tebtebba Foundation.

Toledo, Victor. 2001a. Indigenous peoples and biodiversity. In *The Encyclopedia of Biodiversity,* edited by S. Levin, G. C. Daily, R. K. Colwell, J. Lubchenco, H. A. Mooney, E. D. Schulze, and D. Tilman, 1181–1197. San Diego: Academic Press.

———. 2001b. Biocultural diversity and local power in Mexico. Challenging globalization. In *On Biocultural Diversity: Linking Language, Knowledge, and the Environment,* edited by Luisa Maffi, 474–483. Washington, DC: Smithsonian Institution Press.

———. 2006. *Ecología, espiritualidad, conocimiento: De la sociedad del riesgo a la sociedad sustentable.* Red Utopia, Morelia, Michoacan, Mexico: Jitanjáfora Morelia Editorial.

Toledo, Victor, Pablo Alarcón-Chaire, and Patricia Moguel. 2001. El atlas etnoecológico de México y Centroamérica: Fundamentos, métodos y resultados. *Etnoecológica* 6(8):7–34.

Valladolid, Andres. 2005. Kawsay mama (8). Introducción al monitoreo de la variabilidad de cultivos nativos y sus parientes silvestres. PRATEC (Proyecto Andino para las Tecnologías Campesinas [Andean Project for Peasant Technologies]). Lima, Peru: PRATEC.

Valladolid R., Julio. 2001. Crianza de la Agrobiodiversidad en los Andes del Perú. Kawsay mama (1). PRATEC (Proyecto Andino para las Tecnologías Campesinas [Andean Project for Peasant Technologies]). Lima, Peru: PRATEC.

———. 2005. Importancia de la conservación in situ de la diversidad y variabilidad de las plantas nativas cultivadas y sus parientes silvestres y culturales en la región Andino-Amazónica del Peru. Kawsay mama (9). PRATEC (Proyecto Andino para las Tecnologías Campesinas [Andean Project for Peasant Technologies]). Peru: PRATEC.

Valladolid R., Julio and Fréderique Apffel-Marglin. 2001. Andean cosmovision and the nurturing of biodiversity in the peasant Chacra. In *Indigenous Traditions and Ecology: The Interbeing of Cosmology and Community,* edited by J. Grim, 639–670, Cambridge, MA: Harvard University Press.

Van Dam, Chris. 1999. La tenencia de la tierra en América Latina: El estado del arte de la discusión en la región. Documento especialmente preparado para la Iniciativa Global Tierra, Territorios y Derechos de Acceso. Union Internacional para la Conservacion de la Naturaleza, UICN, Oficina Regional para Sud América.

Varese, Stefano. 1996. Parroquialismo y globalización: Las etnicidades indigenas ante el tercer milenio. In *Pueblos Indios, Soberania y Globalismo*, edited by S. Varese, 15–30. Quito, Ecuador: Ediciones Abya Yala.

——. 2001. The territorial roots of Latin American indigenous peoples' movement for sovereignty. *International Social Science Review* 2(2):201–217.

CHAPTER FIVE

Saving Our Seeds

An Indigenous Perspective from Cotacachi, Ecuador

MAGDALENA FUERES, RODRIGO FLORES,
AND ROSITA RAMOS

As members of the indigenous communities of Cotacachi, Ecuador, we would like to describe our experiences over the past few years with the recovery of our native plants. Through various initiatives implemented by foreign friends and Ecuadorian compatriots, our indigenous organizations—UNORCAC (*Unión de Organizaciones de Campesinas y Indígenas de Cotacachi*) and *Runa Tupari* (an indigenous rural tourism agency)—and the *comunas* (communities) have benefited from interest in and support of our ancestral crops and medicinal plants. We feel that this collaboration has already brought back many of our ancestral plants, which are today growing in our *huertas* (gardens) and *chacras* (fields). As new programs are implemented in Cotacachi to recover our ancestral plants, we want to offer some basic principles on our indigenous view of plants in our culture and why the principles must be respected.

As indigenous Cotacacheños, we are fighting to save and recuperate our way of life in its totality, including our customs, language, territory, and our native plants. We have made progress since 1978, when we first organized together and openly resisted racism, oppression, and other indignities. However, today new threats from both outside and inside our communities endanger our indigenous way of life. In response, we have developed the theme *"Desarrollo con Identidad"* (Development with Identity), which is the guiding principle for the collaborative projects in which we now engage. We see a need to bring to our communities more economic opportunities, improved health, and better education for our children. However, it is also important that we do not lose our identity in achieving this better life. For over five hundred years, through slavery and alienation, we have drawn strength from the traditional ways of our ancestors, and we will continue to do the same in the future. Whether working with government officials, conservation and development workers, or scientists, we feel the manner of collaboration should be based on respect for our culture and our equality as human beings without paternalism.

In this regard, we feel it is important that visitors to our communities understand how indigenous people feel about Mother Earth, or *Pachamama*, as we call her in Quichua. In relation to all of nature, including the sun, moon, mountains, land, plants, and animals, we hold a cosmovision that we believe must be respected when dealing with ancestral plants. We believe that we will survive by cooperating among ourselves and sharing knowledge and the gifts of Pachamama. For us, the plants and animals are all a part of the connections between the *Runa* (indigenous people) and Pachamama. Our cosmovision is integral, guiding all decisions and strategies. For this reason, we do not speak of "conservation" but of coexistence of all living forms. Thus, the diversity of life—of which the Runa are just a part—must be valued, respected, and encouraged. We also want our young people to learn about the plants cultivated in the past, by the ancestors, and to maintain knowledge about these plants, which our elderly community members still possess and continue to practice.

We are also painfully aware that in our lifetime our people have started to lose many of the native crops of Cotacachi. We are likewise losing everyday the knowledge of how to cultivate, prepare, and cook traditional food. When we were children, our parents grew many different kinds of potatoes, maize, choclos, beans, quinoa, zambo, zapallo, other *granos* (grains), and medicinal plants. Today, we plant only a few varieties. The many varieties of our parents have disappeared or are now hard to find. After land reform, the *ingenieros agrónomos* (agronomists) from the Ministry of Agriculture and other organizations came and taught us about new improved varieties and how to obtain and use fertilizers and pesticides. During those years, thieves stole much of our livestock, so today we have little corral manure for our gardens and fields. Each year to get a good harvest we have to put more and more chemical fertilizers and pesticides, which we must purchase from the store, but these are very expensive and now the crops do not yield well. Today, our farmers find it harder and harder to plant as the rains do not come when they should, the springs and rivers are becoming drier, and good seeds are increasingly difficult to find. Many of our young people continue to leave Cotacachi for work in Quito, and this makes it difficult to cultivate the land. Sometimes in our communities, only old people and children go to the fields during the day.

As the elders pass away, we are also losing the ancestral knowledge about the land and the plants. This is a very serious matter for our communities that must be halted and reversed. When we were young, our elders would tell us stories of Mama Cotacachi, our Mother Mountain, who sees everything and approves or disapproves of how people are behaving. The old folks said that we are her children and that Mama Cotacachi provides for us who live all around her *falda*, her skirt. Mama Cotacachi, the old people told us, is not filled with lava and rock like scientists say but is full of grains and potatoes and all of the energy and diversity of all our crops. By respecting Mama Cotacachi, our elders and parents always had good communication with the land, which gave us good harvests. But in the last few years, Mama Cotacachi has not been happy with the people. She is unhappy because we have blackened her beautiful skirt by burning, and the snow adorning her beautiful long hair has gone away. Scientists say it is because of global warming, but we also believe it is because people, especially our young people, no longer have the same respect for Mama Cotacachi as before. The youth are becoming accustomed to new foods, the "light foods" of the mestizos, and no longer want to eat our traditional foods made of Andean granos. We believe that one way to recover respect for our culture is to teach our youth why our plants are valuable, why caring for a plant is sacred, and how the plants nourish and cure us.

We are proud of three activities in which we have participated as indigenous leaders in restoring our traditional plants. These activities respected our indigenous culture while recuperating seeds, advancing the education of our youth, and helping to increase the financial status of our people.

Creating a "Memory Bank" of Plant Knowledge

We collaborated with Virginia D. Nazarea and Maricel Piniero, anthropologists from the University of Georgia, to document and preserve our elders' knowledge associated with agricultural crops and medicinal plants so that neither seeds nor knowledge will be lost (see figure 5.1). With a small annual fund, twenty U.S. dollars provided a scholarship for matriculation fees and books for each of the students whose families were willing to help the children record the elders' knowledge about

FIGURE 5.1. Memory banking youth of Cotacachi. Virginia D. Nazarea and Maricel Piniero are in the center. Photo by Robert E. Rhoades.

plants. Between 1999 and 2006, fifteen indigenous children annually received the scholarship and have written reports as part of their school activities. They asked their parents and their grandparents a series of basic questions, such as "What varieties did you grow when you were young?" and "How did farming change over time?" The young memory-banking scholars (*becarios*) also collected seeds from their families and communities and planted them in a communal biodiversity garden on the grounds of Jambi Mascaric (Search for Health) and in their community schools. They also created public exhibits and posters of the plants and elders' knowledge during special days at the school.

We also gathered in *Circulos de Ancianos* (Circles of Elders) at the church and in Jambi Mascaric and asked participants to tell stories about the plants they knew as young people and to compare them with what they grow today. This information was recorded and kept in our offices in Jambi Mascaric for use by the communities and other projects interested in recovery of our ancestral crops. A main priority of our indigenous families, communities, and organizations is the education of future leaders, particularly young women. Two of the memory-banking scholars have now finished college, one with a degree in education and the other with a degree in music. We feel the memory-banking scholarship program, which supports young people in school while they help recover the knowledge and seeds of the ancestors, is a practical way forward.

Establishing a Farm of the Ancestral Futures

In 2002, we established our own indigenous seed farm, in collaboration with Robert E. Rhoades and Shiloh Moates, also from the University of Georgia. The *Finca de Futuros Ancestrales* (Farm of the Ancestral Futures) serves to teach our youth about the old crops. Its goal was to help recuperate Andean crops that have been mostly or completely abandoned. Coauthor Magdalena Fueres donated the two hectares of land on her family farm in Ushugpungo located in the high zone of Cotacachi. In reciprocity, the same families of the memory bankers described above provided labor, seeds, and knowledge. The mothers (one father participated) provided two types of quinoa, peas, chocos, habas, mellocos, chaucha potato, mashuas, and ocas, all from their own farms. Shiloh Moates also helped to compose an agreement with the *Instituto Nacional Autonomo de Investigaciones Agropecuarias* (INIAP) and its national genebank located outside Quito to provide seeds of endangered indigenous crops historically grown in Cotacachi. Among more than sixty species traditionally cultivated in Cotacachi, we were able to obtain two dozen species for recuperation and reintegration. The ancestral farm was a place that encouraged a participatory dialogue among indigenous young people, their elders, and scientists on the issue of loss and recovery of plants. For us, bringing back our old crops was symbolic of the recovery of our traditions.

While the children were attending school during the day, we organized with the parents a weekly *minga* for planting, weeding, cultivating, and harvesting (see figure 5.2). A minga is a communal work party to mobilize labor and is also a way of demonstrating community and showing reciprocity and solidarity among ourselves. We believe that having a minga is what makes us indigenous and sets us apart from the mestizos.

The farm was divided into halves. One side was dedicated to local varieties and planted according to the parents' directions, while the other half was planted with INIAP varieties using the scientists' specifications on spacing and planting depth. INIAP provided twenty-five varieties each of oca and mashua, as well as four varieties of achira in exchange for the rights to visit the plot and to receive information about how the donated varieties performed. Our local side was laid out according to the way we plant in our fields and gardens. In keeping with traditional practices, the farm was entirely organic and managed using local knowledge. The parents led this process by teaching how to plant "ally" crops together and keeping well apart the crops that "don't get along," just as they had learned from their parents. The Farm of the Ancestral Futures was one of the first projects in Cotacachi to have as its goal the connection of

FIGURE 5.2. Planting the Farm of the Ancestral Futures. Photo by Virginia D. Nazarea.

our culture with the ancestral crops. Since then, many programs have come to Cotacachi, and we have tried to incorporate these same values into their design.

Incorporating Ancestral Plants into Rural Tourism Initiatives

For over twenty years, the indigenous communities of Ecuador have been searching for alternatives in order to adapt to the new social order and to confront our economic problems. We no longer can survive by working only the land, and we feel that the trend of young people migrating to cities or foreign countries in search of work is not a long-term solution. We feel that a more sustainable solution is community-based tourism, which draws its strength from our own cultural uniqueness. To this end, the *Federación Plurinacional de Turismo Comunitario del Ecuador* (National Intercultural Federation of Community-based Tourism of Ecuador) was established with fifty-seven members from different communities in the Amazon, the Andes, and the coast. The aim of the program is to show the diversity and distinctiveness of culture in our country while gaining income for our communities. Following the principles of "development with identity," we would like to see tourism come to our villages in a way that respects and supports our cosmovision and culture.

In Cotacachi, the UNORCAC, together with our indigenous communities, created the rural tourism agency Runa Tupari, which in Quichua means "meet indigenous people." Through this program, we established a tour operating agency in nearby Otavalo where tourists enter the region, constructed rural lodges in four Cotacachi villages, trained community leaders and guides, and developed a marketing program to attract tourists to rural areas. In collaboration with INIAP-UNORCAC, we established near each of the rural lodges an "indigenous identity and agrobiodiversity garden" (see figure 5.3). These gardens contain native plants to help educate tourists, and they also serve to increase local plant diversity. We found that the experience of tourists is enriched if they can see how we grow our traditional crops and prepare our native dishes. Other aspects of our culture and diet, such as the role of the *cuy* (guinea pig), are also of keen interest to tourists.

We realize that our native crops are in competition for the hearts of our people with mestizo foods like bread and rice that today are easy to prepare and readily available in the local market. Our native grains and tubers require special preparations that take time. We need better

FIGURE 5.3. Sign above rural lodge stating "Rural Tourism: Indigenous Identity and Agrobiodiversity, Cotacachi, Ecuador." Photo by Robert E. Rhoades.

markets and profits to continue planting our native crops. We hope to increase our production while displaying the great diversity of shapes, colors, and unique flavors of our native species. Agrotourism is one way we can rescue and revalidate the cultural and culinary aspects of Cotacachi's native crops. One of the tours we offer is an agroculinary tour, in which tourists take native meals in the rural lodges and participate in a guided tour of the homegarden near their lodge. In this manner, tourists visiting Cotacachi learn our rich agricultural patrimony and cuisine. They can experience the rhythm of work and life in the villages and appreciate how our agricultural calendar is built around plants and animals. They will observe the seasons that influence our diet and cooking, the changing colors of the landscape, and the activities in our communities. Today, tourists from more than seventeen countries have stayed in the rural lodges, and we expect even more visitors in the years to come.

To conclude, we hope that we have provided some insight into the principles that we feel are important in the recovery and future nourishment of the plants and animals we have in Cotacachi, along with some examples of our activities based on these principles. We support development (and conservation) if these measures do not cause us to forget who we are in the process. Our survival as indigenous people is built on working together in the minga. All living things are in coexistence as a part of Pachamama, including Runa, the people, and the plants and animals. Everything we do must consider and respect this connection. Finally, we must build a bridge between our youth and our elders, who still hold the knowledge. In our way of thinking, the path to the future is through the wisdom of our ancestors. This is the path we must take to save and make bountiful the diversity of our plants and animals.

CHAPTER SIX

People, Place, and Plants in the Pacific Coast of Colombia

JUANA CAMACHO

Across time and space women have played important roles in plant management and *in situ* conservation. In the past years there has been increasing recognition of the significance of women's agricultural practices, particularly in homegardens (Oakley and Momsen 2007; Aguilar-Stoen, Moe, and Camargo-Ricalde 2009). Women's gendered environmental knowledge and gardening skills offer multiple economic, nutritional, health, and livelihood diversification benefits (Rocheleau et al. 1997; Howard 2003). Anthropological and geographical conceptualizations of place have focused on the intersection of natural and cultural processes, global-local interactions, power relations, and identity (Gupta and Ferguson 1997; Escobar 2001; Escobar et al. 2002; Raffles 2002). How places are rendered meaningful by individuals and groups and the complex ways in which they anchor lives in localized social formations have produced textured cultural understandings of emplacement, dwelling, and uprooting (Feld and Basso 1996). In this regard, feminist scholars have called attention to the gendered dimension of place and women's centrality to the social production and experience of place and place making (Domosh 1998; McDowell 1999). Through homegardening, women not only create the conditions for the growth and development of plants but also effect emplacement through their creative and nurturing practices. Virginia D. Nazarea (2005) notes that homegardens operate as repositories of cultural memory and attest to a people's history, agricultural philosophies, and emotions. Environmental transformation, territorial displacement, and cultural dislocation are physically and psychologically traumatic experiences. The loss of home and familiar identity referents leaves people disoriented, insecure, in pain, and without a sense of place or belonging. For rural and agricultural communities forced out of their lands, loss of connections with plants and place makes reconstruction of life in urban settings even more distressing.

This chapter explores Afro-Colombian women's homeplace making in the Colombian Pacific coast through an examination of their environmental and homegarden management practices. I argue that in a context of indentured slavery, adaptation to the tropical rainforest, and structural exclusion, black women's environmental knowledge and social practices have contributed to the reconstruction of the social and cultural fabric and to the defense of place and culture. Yet their social and environmental knowledge and practices remain invisible to the public eye. In the past decades, armed conflict for control over this geostrategic territory by paramilitary, guerrilla, and state forces has led to systematic human rights violations and massive internal displacement of local populations. Local economies and livelihoods are further affected by habitat destruction and fragmentation caused by colonization and extractive activities. Afro-Colombians make up a disproportionate number of displaced people that have been forced to abandon their land and belongings. Displaced and uprooted individuals and families have sought refuge in urban centers without guarantee of a safe return to their communities. Forced displacement means breaking away from two constitutive elements of their identity and historical experience: the collectivity and the territory (Jimeno 2001). Material and moral dispossession means loss of the means of subsistence and the sense of belonging. Women, who are still the primary caretakers of children, have been particularly affected by the loss of land, resources, and social networks that sustained their livelihoods, social relations, and identities. Women have faced the challenge of making place in new settings where they are discriminated against not only on the basis of race, gender, and place of origin but also as stigmatized *desplazadas* (displaced people). In what follows, I attempt to provide a more complex and textured understanding of the situated knowledges and practices of black women, with particular attention to their agency in maneuvering within the constraints of structural forces and conjunctural challenges.[1]

Homeplace making is the process of physical and symbolic occupation, appropriation, and transformation of the natural surroundings into a familiar and intimate place for the production and reproduction of Afro-Colombian culture. Based on several years of work in the Colombian Pacific coast (Camacho 1999, 2000), I argue that for Afro-Colombians the experience of place is a material ecological reality, a network of social relations and practices, and a set of intersecting meanings and memories.

Their notion of place is intimately tied to a long history of deterritorialization from Africa, emplacement in the tropical rain forest, and resistance to capitalist expansion, armed conflict, and massive displacement in the region. Homeplace is not confined to the domestic feminine domain but encompasses a larger sociospatial reality that is at the heart of current Afro-Colombian political struggles for the defense of place. Drawing on bell hooks's (1990) discussion of homeplace among African Americans, I highlight the importance of black women's everyday work in a context of structural, socioeconomic, racial, and gender discrimination and exclusion. In view of sexist, racist oppression and segregation in the United States, hooks notes that as homes and communities became spaces that enabled black people to be dignified and valued subjects, black women's conventional homeplace-making role acquired a politically subversive dimension. Women's active contribution to the reconstruction and "remembering" of fractured identities and subjectivities made possible the emergence of communities of solidarity and the contemporary repolitization of homeplace making.

The gendered relation of Afro-Colombians with different spaces or areas of their territory is expressed in women's multiple roles in rituals of embodiment, emplacement, and identity construction. Within this context, homegarden management and *in situ* conservation exemplify the spatialization of social and ecological relations beyond the domestic realm. Here, I note some of the implications of civil war and forced displacement for black women in regard to their livelihoods, social relations, and identity. I use the terms black and Afro-Colombian interchangeably; black is the most common term employed to refer to people of African descent, but the term Afro-Colombian has gained prominence in the past two decades as part of an emerging politics of difference, in which black communities adopted an ethnic mark and became subject to special territorial and cultural rights as an ethnic minority. At present, rural inhabitants continue to employ the term "black," while "Afro-Colombian" is more prevalent among urban political and cultural leaders along with activists, ethnic organizations, academics, and state institutions. A more thorough discussion of the ethnicization of blackness and the politization of cultural and ethnic difference can be found elsewhere (e.g., Restrepo 2001; Escobar 2008; Asher 2009).

The Territorialization of Culture: Afro-Colombian Struggle *To Be* in Place

Recent conceptualizations of space and place point to the mutual constitution of social and spatial processes, also known as the sociospatial dialectic (Soja 1980). Scholars agree that space is not a neutral physical area but is turned into place by the social relations, practices, identities, and meanings that occur in a geographical location. Place is socially and heterogeneously constructed according to people's class, gender, ethnicity, age, or ecological experience. Social relations are expressed in the way individuals and collectivities appropriate and define space, its boundaries, and who and what is socially and spatially included or excluded. Henri Lefebvre notes that "socio-political contradictions are realized spatially [and] spatial contradictions 'express' conflicts between socio-political interests and forces" (1991:365). In the Colombian Pacific, perceptions and experiences of space and place are intimately related to the process of social and ecological adaptation to and appropriation of tropical rainforest environments by Afro-Colombians who historically have been socioeconomically and geographically marginalized by Andean white and mestizo powers.[2]

For the past two decades, Afro-Colombian populations of the Pacific coast have been engaged in a systematic process of social organization and political mobilization for the right To Be, or to exist as citizens with full rights. As the Afro-Colombian social movement emerged, it structured its claims of national ethnic and place-based political mobilizations that led to the 1991 constitutional recognition of the multiethnic and pluricultural nature of the Colombian nation. Drawing on the historical and political experience of indigenous peoples and the International Labor Organization covenant on minority rights (Wouters 2001), and with support of the Catholic Church, activists in the black movement engaged in a new politics of representation and of difference. Blacks represented themselves as an ethnic minority and posited ethnic rights, cultural identity, territory, and autonomy as the pillars of their struggle. Cultural identity was defined on the basis of a shared experience of enslavement and incorporation into colonial socioeconomic structures, a common form of social organization centered on the extended family and kinship networks, a close interaction with nature, and a rich oral and musical tradition. The territory over which they claimed sovereignty was viewed as the concrete material space where their distinct culture and livelihoods were created and recreated upon the end of slavery in 1851. Although a large number of Afro-Colombians live in urban areas, the black movement focused primarily on rural populations settled on the Pacific coast—a large swath of rainforest that extends between Ecuador, Colombia, and Panama, and is one of the world's top twenty-five biodiversity hot spots (figure 6.1).

The Pacific region is complex due to not only its biological and cultural diversity but also the diverse, and contradictory socioeconomic, political,

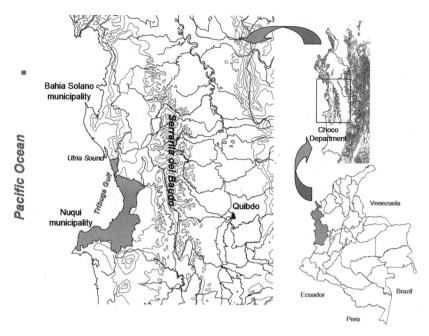

FIGURE 6.1. Map of the Colombian Pacific region.

and cultural processes that are taking place as a result of state and private initiatives (Escobar and Pedrosa 1996; Flores and Millan 2007; Rapoport Center 2007; Escobar 2008). The region includes a variety of wet and moist forests, mangrove forests, beaches, rocky shorelines, and small mountain systems. It is considered a priority for conservation due to its considerable level of endemic plant species, birds, and amphibians (Alberico 1993; Gentry 1993). The Pacific coastal region is also home to indigenous and Afro-Colombian communities whose livelihoods depend directly on natural resources. Indigenous communities tend to be in the interior of the forest close to riverheads, while black settlements are nearer to the coast. Most mestizo immigrants from the country's interior are settled in the region's urban centers. Populations of the Pacific coast are experiencing the social and environmental effects of accelerated land acquisition by outsiders for extensive cattle breeding, agroindustrial projects, and tourist ventures, as well as unregulated fishing, timber extraction, mining, and cultivation of coca. More recently, the region has seen the expansion of extensive oil palm plantations at the expense of the forest and the radical transformation of local productive systems. Oil palm cultivation is presented as the area's new rural modernization panacea, given the international demand for agrofuels. The area also offers a strategic location for the establishment of corridors where coca and arms circulate between the Pacific Ocean and the interior of the country.

While large-scale development projects are being implemented, environmentalist organizations, social and ethnic movements, and some state agencies have mounted regional sustainable development initiatives and projects for the protection of common lands and people's lives. Violent conflict among different armed sectors over control of the territory and its natural resources has forced people to leave and abandon their lands. Today, despite having one of the most ethnically progressive legislations in the world on their side, indigenous and Afro-Colombian communities have been disproportionately affected by the hostile dynamics of guerrillas, paramilitary forces, and the national army against unarmed civilians. Government response not only has been insufficient and inadequate to guarantee basic human rights but also, in some cases, has contributed to further military repression, stigmatization, and exclusion.

Afro-Colombians are the descendants of enslaved Africans brought to Colombia in the seventeenth century to work in mining and plantations after the country experienced massive indigenous depopulation brought about by disease, famine, warfare, and overexploitation. Since slavery was abolished, freed slaves working in the mining camps in the Pacific lowlands moved in all directions, colonizing rivers and founding coastal villages in a discontinuous settlement pattern. The new settlers developed subsistence economies based on seasonal exploitation of the different ecosystems (forests, rivers, mangroves, coasts) for hunting, gathering, mining, fishing, and farming. Local populations remained tied to national and international markets in sporadic extractive booms such as rubber, vegetable, ivory, timber, and mangrove bark, and more recently tourism, commerce, drug smuggling, and wage labor. Although the natural richness of the Pacific has enabled the accumulation of wealth in national Andean centers of power, the area has remained socioeconomically marginal and has among the highest poverty, child mortality and morbidity levels, and the lowest social security coverage, in the country (Afrodes—Global Rights 2008).

The Afro-Colombian experience of adaptation and modification of the complex terrestrial and aquatic ecosystems illustrates the transformation of space into place. Place is defined here as "the ensemble of relations and practices between the natural and the social worlds at the level of body, home, habitat, and community. In place, we thus find a convergence of nature and culture, women and environment, ethnicity and ecology" (Escobar et al. 2002:29). Place in the Colombian Pacific is made up of different sites or places whose appropriation, use, and management are based on a spatial and gendered division of labor. Sites are not only for productive activities but also for entertainment, play, and ritual. Indeed, as summarized by black women activists (Grueso and Arroyo 2002), place is where the social matrix is woven generation after generation.

Historically, this area, known as the Pacific Chocó, has been designated a national forest reserve, although it is home to rural and urban Afro-Colombian communities along with other indigenous groups and mestizo settlements. As early as the 1980s, indigenous reserves were established and legalized based on ancestral rights claims and arguments about the harmonious indigenous-environmental relationship. However, blacks were regarded as uneducated settlers without culture and therefore without any legitimate claim to the land. In the 1990s, the emerging black movement, with the support of indigenous leaders and mestizo collaborators, initiated claims to territorial rights and substantiated them with arguments about the sustainability of their productive practices and rich cultural traditions. In 1993, a progressive legislative measure granted special cultural and collective land rights to Afro-Colombians as an ethnic minority. In a short period, blacks moved from historical invisibility at the margins to the center stage of national ethnic politics (Asher 2009).

The black movement's demands for ethnic and territorial recognition against the interests of the state, private investors, and mestizo colonists have defined the concept of territory as the bounded site of ethnic sovereignty. The cultural, geopolitical, and biogeographic conceptualization of territory and territoriality, and territory-region (Proceso de Comunidades Negras 1996–1998), as the autonomous place for the reproduction of a distinct culture has been instrumental to frame and sustain Afro-Colombian political demands. The use of the concept of territory, however, often obscures the complex, multiple, affective, and gendered relationship of black communities with homeplace. A more comprehensive perspective has been proposed by the women from the Network of Black Women of the Pacific, who assert that

> First we want for all black women to harbor the Pacific in their hearts and minds, that is, for all women to have the political awareness of the ethno-cultural and territorial rights. . . . Second, we wish for all black women to understand and embrace the problematic aspirations of the Black Community as a people, and for us to express and mobilize our needs and interests as black women from within the Black Community. (Grueso and Arroyo 2002:63)

For Afro-Colombians, the construction and experience of homeplace as a material ecological reality, a network of social relations and practices, and a set of intersecting meanings and memories are the result of a complex history of deterritorialization, emplacement, and resistance.

Gendering Space

Feminist geographers have argued that space is not gender neutral and that social asymmetries are expressed in the way space is organized and segregated for inclusion or exclusion (Pollock 1988; Rose 1993; McDowell 1999). Although their work is largely focused on the analysis of productive/public and reproductive/private spheres in modern Western societies, they trace this dichotomous separation as a particular development of the historical process of industrial capitalism and urban bourgeois domesticity. They further note that social asymmetries are expressed in the social and economic devaluation of the feminized reproductive/private realm.

The Afro-Colombian gendering of space in Chocó is a combination of local concepts of nature and gendered material practices. The sociospatial organization of labor rests on the social, productive, and spatial complementarity of men and women and the positive valuation of each gender's roles. The primary unit of social organization and work is the family, and agriculture is the main productive activity around which all other endeavors revolve. At times of labor shortages, cooperative and reciprocal work forms such as *minga* (collective work) and *cambio de mano* (reciprocity among individuals) are used. Most work activities are performed collectively by the entire family, though others are exclusively masculine and/or feminine. Women take part in agriculture, fishing, markets, and trade in addition to domestic tasks in spaces surrounding the house. With the exception of the privately owned agricultural lands and village lots, most areas are commons or public lands used interchangeably by men and women. Customary property rights are passed in a bilateral pattern of inheritance; men and women have the same inheritance rights over family possessions. Land rights are also acquired through the investment of labor and purchase.

In previous research (Camacho and Tapia 1997) on the sociospatial organization of labor, we identified eight major activity areas: house, village, river, beach, sea, mangrove, wild forest, and tame forest, which have a rough correspondence with local ecosystems and are classified according to social and ecological criteria (table 6.1). The main classificatory criterion is the degree of human intervention, which influences vegetation cover, the presence of wild animals and other nonhuman beings, and the type and intensity of social activity. The activity areas can be arranged along a flexible continuum from wild/masculine/untamed to tame/feminine/domestic. Eduardo Restrepo (1996) found a similar classificatory scheme or "environmental grammar" among other Afro-Colombian populations. Wild or primary forest (*monte bravo*) is charac-

Table 6.1. Local classifications and use of spaces in territories of black communities

Area	General Characteristics	Use	Gender Activities
Primary forest, *monte bravo*	Wild and dangerous, home to beasts and poisonous animals. Men "tame" it with their work by felling trees and clearing the land. It is cool because of forest canopy cover.	Hunting, extraction of timber, palms, lianas, and medicinal plants	Mostly male
Intervened "tame" forest, *monte amansado, monte biche*	Fallow forest, agricultural fields, productive land.	Agriculture, hunting, gathering of fruits, plants, palms; collection of ant soil	Male and female
River	Communicates the village and the agricultural parcels. The name of the river and the village is generally the same. Local people self-define according to the river name.	Transportation, fishing, gathering mollusks, extraction of sand and rocks, washing clothes, swimming and entertainment	Male and female
Mangrove	Cool, swampy, and damp place. Pregnant women or during the menstrual period can catch cold and become ill.	Fishing, gathering firewood, crabs, clams	Male and female
Sea	Calm on the surface and the shore but wild and treacherous. Cannot be tamed. Occasionally people swim in the sea.	Transportation, fishing, entertainment	Male

(*continued*)

Table 6.1. Local classifications and use of spaces in territories of black communities (continued)

Area	General Characteristics	Use	Gender Activities
Beach	Main terrestrial means of communication between coastal villages.	Gathering clams, shells for crafts, decomposing leaves and driftwood for zoteas, sand and rocks for construction; places for entertainment	Male and female
Village	A humanized and tame place. Danger arises from human action.	A place for living, socializing, and working; site for government institutions, pubic services, commerce	Male and female
House	A humanized and feminine space. Provides sense of place and belonging.	Domestic work, eating, socialization, rest, hold wakes; grow plants in patios and zoteas, tend chickens and pigs	Mostly female

terized by thick vegetation and is home to wild beasts, dangerous and poisonous animals or *avichuchos*, and nonhuman beings. This is a male area par excellence used for hunting, timber extraction, and gathering of potent medicinal plants and where women go once men "tame" it by clearing it for agriculture. Taming is possible because men are endowed with the physical strength necessary to undertake the arduous and risky task of clearing a perceived dangerous and wild place.[3] According to local testimonies men belong in the forest, men's work is in the forest, felling the monte bravo is men's work and the wild forest must be tamed by men. "Tame" refers to a condition resulting from human intervention, which makes possible the socialization of space through productive labor. When the primary forest becomes a *monte amansado* (tame forest), it becomes a safe space for women and children, who participate in the planting and harvesting of crops. The river, the beach, the sea, and the mangrove are spaces for labor and entertainment for men and women and whose appropriation involves human presence and work.

The village and the house are at the other end of the spectrum and symbolize the main areas of human intervention and activity. The village is perceived as the place for socialization, wage work, and institutional interaction with various state and private actors. According to one local inhabitant, "the town is for everyone, it is said that men belong in the street because they go out and drink and fight but women also go in the street to chat, run errands, and dance. The home however, is mostly female: there is the kitchen, the hearth and the garden." The home, in turn, is a humanized and feminine space and a center of biological and social reproduction. In the Chocó coast, the home is a space that synthesizes the social, ecological, economic, and cultural relationships of Afro-Colombians with the larger areas of the territory. The home is where material and energy flows converge in the form of crops, foods, plants, and building materials that are brought to the home from the sea, the river, the mangrove, the forest, and the village to be processed, transformed, consumed, or stored.

The home defines women's economic and social roles and identities in the organization and reproduction of the family and the kinship group. The historical value of black women's socioeconomic role, their centrality in social organization and in the transmission of cultural patterns since their introduction as slaves in the seventeenth century, has been widely acknowledged in Afro-Colombian scholarship. Some authors even characterize Afro-Colombian family and social organization as matrifocal and matrilineal given the historical consolidation of lineages and kinship ties around African women who were forced to have multiple sexual partners to reproduce slave labor and whose children were identified through maternal descent (see Camacho 2004 for an overview). Although the heterogeneity of black social organization and family structure has been increasingly recognized, the importance of women's domestic role is undisputed.

In domestic spaces, culture and memories are transmitted from elders and mothers to children through teaching by example and enskillment (Palsson 1994; Ingold 2000). Enskillment is the process by which capabilities of action and perception are acquired through training and experience in the performance of particular tasks within a community of practice. Daily engagement with the natural surroundings contributes to the development of skills as well as certain perceptual and sensory sensitivities that constitute people's embodied knowledge. Homes are also places where women sustain life, symbolically and materially, from beginning to end: childbirth, daily preparation of meals, various healing practices, and the performance of funerary rituals.

The cultivation of a wide variety of plants is also an important domestic task because plants figure prominently in different settings for distinct purposes: food, medicine, ornament, fodder, witchcraft, and ritual celebrations. Women are the primary managers of seeds, medicinal plants, and domestic crops. Their gardening practices are fundamental to the process of homeplace making for they not only provide local livelihoods and promote social well-being but also constitute means by which the landscape is humanized and transformed in subtle but meaningful ways.

Homegardens and Zoteas

Homegardens are microenvironments within larger agricultural systems that concentrate in a relatively small space different species with diverse economic, nutritional, and medicinal functions (Eyzaguirre and Linares 2004). In the Pacific Chocó, homegardens include a wide variety of native and introduced trees, shrubs, and herbs in heterogeneous arrangements, which contribute to *in situ* agrobiodiversity maintenance and household well-being. In the spaces surrounding the home—the front garden, the back patio, and the raised garden, or *zotea*—women cultivate an impressive array of medicinal, culinary, ornamental, and "power plants" of social, cultural, and biological relevance (tables 6.2, 6.3). Species grown in domestic spaces complement the agricultural crops cultivated in family fields and the wild plants from the primary forest but management practices differ for each. Gardens contain more species diversity but fewer numbers of individual plants, which, according to local women, must be cared for as if they were family. In a socieconomically marginal place such as the coastal Chocó, this diversity is strategic for broadening the household productive, economic, medicinal, and nutritional range of options. Sound management of this local diversity has enabled local populations to be autonomous and self-sufficient until recent times.

For Afro-Colombians, garden management is a feminine task that begins during childhood in the family home and is reinforced when a woman signals her adult status by establishing a family in a separate house with her own garden. An abundant garden is a source of pride and prestige for women and is a vehicle for engaging in a productive network of exchanges, in which plants, coupled with a vast knowledge accumulated through oral tradition, circulate among family members and neighbors. Plants have cultural and social meanings not only because of their everyday use but also because they are identity markers for Afro-Colombian women who possess a tradition of plant use and plant lore. This knowledge is passed from one generation to the next through medicinal preparations, culinary traditions, stories, songs, and jokes. Plants cannot be divorced from the

Table 6.2. Plant species grown in patios and zoteas

Family	Species	Habit	Origin	Use
Acanthaceae				
	Acanthaceae	Herb	?	Medicine
	Justicia sp.	Herb	Native	Medicine
	Sanchezia sp.	Herb	Native	Medicine
	Trichantera gigantea	Tree	Native	Fodder, shade, wooden posts, live fences
Alliaceae				
	Allium fistulosum	Herb	Introduced	Food, medicine
	Allium sp.	Herb	Introduced	Food, medicine
Amaranthaceae				
	Alternanthera sp.	Herb	Native	Medicine
	Amaranthaceae	Herb	?	Medicine
	Cf. *Iresine* sp.	Bush	Native	Fruit, shade
Anacardiaceae				
	Anacardium occidentalis L.	Tree	Native	Fruit, shade
	Mangifera indica L.	Tree	Introduced	Fruit, shade, medicine
	Spondias purpurea L.	Tree	Introduced	Fruit, shade
Annonaceae				
	Annona cherimolia Mill.	Tree	Introduced	Fruit, shade
	Annona muricata L.	Tree	Introduced	Fruit, shade
	Annona sp.	Tree	Native	Fruit, shade
	Annona squamosa L.	Tree	Native	Fruit, shade
Apiaceace				
	Eryngium foetidum L.	Herb	Native	Food, medicine
Apocynaceae				
	Tabernaemontana sp.	Bush	Native	Medicine
Araceae				
	Anthurium sp.	Herb	Native	Ornamental

(*continued*)

Table 6.2. Plant species grown in patios and zoteas (continued)

Family	Species	Habit	Origin	Use
	Caladium	Herb	Native	House protection, ornamental
	Rhodospatha sp.	Brush	Native	Live fence
	Xanthosoma spp. 1–3	Herb	Native	Root used for food, leaves used for fodder and wrapping food
Araliaceae				
	Aralia sp.	Herb	Introduced	Household protection, good luck and wealth, ornamental
	Polyscias filicifolia (Moore)	Herb	Introduced	Household protection, good luck and wealth, ornamental
	Polyscias valfoureana	Bush	Introduced	Household protection, good luck and wealth, ornamental
Arecaceae				
	Bactris gasipaes HBK	Palm	Native	Fruit
	Cocos nucifera	Palm	Introduced	Fruit, food, medicine
Asteraceae				
	Ageratum conyzoides L.	Herb	Native	Medicine, charms
	Porophyllum ruderale (Jacq.) Cass	Herb	Native	Medicine
	Porophyllum sp.	Herb	Introduced	Medicine
	Porophyllum sp. 1	Herb	?	Medicine
	Pseudelephanthopus spicatus Rohr	Herb	Native	Medicine
	Tagetes patula L.	Herb	Introduced	Medicine
	Wedelia trilobata (L) A. Hitchc.	Herb	Native	Medicine

Table 6.2. Plant species grown in patios and zoteas (continued)

Family	Species	Habit	Origin	Use
Bignoniaceae				
	Crescentia cujete L.	Tree	Native	Live fence, shade, medicine, fruits are used as gourds
Bixaceae				
	Bixa orellana L.	Bush	Native	Food, medicine
Bombacaceae				
	Quaribea sp.	Tree	Native	Shade, fruit
Brassicaceae				
	Brassica oleracea L.	Herb	Introduced	Food
Bromeliaceae				
	Ananas cosmosus	Herb	Introduced	Fruit, medicine
Cactaceae				
	Opuntia sp.	Bush	Native	Charms
Caesalpinaceae				
	Senna alata L.	Tree	?	Medicine
Cannaceae				
	Canna coccinea Miller	Herb	Introduced	Leaves used for fodder and wrapping food
Caricaceae				
	Carica papaya L.	Bush	Native	Fruit, medicine
Cecropiaceae				
	Cecropia sp.	Tree	Native	Shade, fuel
Chenopodiaceae				
	Chenopodium paniculatum Hook.	Herb	Introduced	Medicine
Combretaceae				
	Terminalia catappa L.	Tree	Native	Shade, fruit, medicine

(*continued*)

Table 6.2. Plant species grown in patios and zoteas (continued)

Family	Species	Habit	Origin	Use
Commelinaceae				
	Commelina spp. 1 and 2	Herb	Native	Medicine, charms
Crassulaceae				
	Kalanchoe sp.	Herb	Introduced	Ornamental, medicine
Cucurbitaceae				
	Cyclanthera pedata Schrad.	Vine	Introduced	Food, medicine
	Momordica charantia L.	Vine	Native	Medicinal
Cyclanthaceae				
	Carludovica palmate R and P	Palm	Native	Fiber for baskets, mats and crafts
Dioscoreaceae				
	Dioscorea alata L.	Ñame Root	Introduced	Food, fodder
Euphorbiaceae				
	Acalypha sp.	Vine	Native	Medicinal, ornamental
	Codiaeum variegatum Blume	Bush	Introduced	Ornamental
	Manihot sculenta Crantz	Bush	Native	Food, fodder
	Phyllanthus niruri L.	Herb	Introduced	Medicine
	Ricinus communis L.	Bush	Introduced	Medicinal
Fabaceae				
	Erythrina edulis Posada	Tree	Native	Shade
	Gliricidia sepium (Jacq.) Wolp	Tree	Native	Live fence, fodder, medicine, fuel
	Phaeolus vulgaris L.	Climbing	Introduced	Food, fodder
Gesneriaceae				
	Drymonia sp.	Vine	Native	Medicine

130

Table 6.2. Plant species grown in patios and zoteas (continued)

Family	Species	Habit	Origin	Use
Hydrangeaceae				
	Hydrangea macrophyla (Thunb.)	Herb	Introduced	Ornamental
Iridiaceae				
	Gladiolus sp.	Herb	Introduced	Ornamental
	Iridiaceae	Herb	Introduced	Charms, love preparations
Lamiaceae				
	Melissa officinalis L.	Herb	Introduced	Medicine
	Mentha rotundifolia (L.) Huds.	Herb	Introduced	Medicine
	Mentha sp.	Herb	Introduced	Medicine
	Ocimum basilicum L.	Herb	Introduced	Medicine, food (seasoning)
	Ocimum micranthum Willd.	Herb	Native	Medicine, food
	Origanum vulgare L.	Herb	Introduced	Food, medicine
	Satureja brownei	Herb	Introduced	Food, medicine
Lauraceae				
	Persea americana Miller	Tree	Native	Shade, fruit, medicine, fuel
Lilliaceae				
	Aloe vera	Herb	Introduced	Medicine, good luck
	Cordyline terminalis	Bush	Introduced	House protection, ornamental, crafts
Loranthaceae				
	Cf. *Struthanthus* sp.	Parasitic	?	Medicine
Lythraceae				
	Lafoensia indica	Bush	Native	Ornamental

(continued)

Table 6.2. Plant species grown in patios and zoteas (continued)

Family	Species	Habit	Origin	Use
Malpighiaceae				
	Bunchosia sp.	Bush	Native	Fruit
Malvaceae				
	Gossipium hirstum L.	Bush	Introduced	House protection, ornamental, medicine
	Hibiscus rosa-sinensis	Bush	Native	Ornamental, medicine
	Hibiscus tiliaceus L.	Tree	Native	Shade, fuel, medicine
	Pavonia sp.	Herb	Native	Medicine
	Sida rhombifolia L.	Herb	Native	Medicine
Maranthaceae				
	Calathea sp.	Herb	Native	House protection
Melastomataceae				
	Bellucia axinanthera Tr.	Bush	Native	Fodder
Mimosaceae				
	Callandra inaequilatera	Tree	Native	Shade, fuel
	Inga spp. 1 and 2	Tree	Native	Shade, fruit, fuel
Moraceae				
	Artocarpus altilis (Z) Fosb.	Tree	Introduced	Shade, fruit, fuel
Museceae				
	Musa AAB	Herb	Introduced	Food, fodder, medicine, leaves used for wrapping
	Musa ABB1	Herb	Introduced	Food, fodder, medicine, leaves used for wrapping

Table 6.2. Plant species grown in patios and zoteas (continued)

Family	Species	Habit	Origin	Use
	Musa ABB2	Herb	Introduced	Food, fodder, medicine, leaves used for wrapping
	Musa ABB3	Herb	Introduced	Food, fodder, medicine, leaves used for wrapping
Myrtaceae				
	Eugenia jambos L.	Tree	Introduced	Shade, fruit, fuel
	Eugenia malaccensis L.	Tree	Introduced	Shade, fruit, fuel
	Psidium guajava L.	Tree	Native	Shade, fruit, fuel, medicine
Oenotheraceae				
	Ludwigia sp.	Herb	native	Soil cover
Passifloraceae				
	Passiflora edulis L.	Vine	Introduced	Fruit
	Passiflora quadrangularis L.	Vine	Native	Fruit
Phytolacaeae				
	Petiveria alliacea	Herb	Native	Household protection, medicine
	Phytolaca sp.	Herb	Native	Food, medicine
Piperaceae				
	Peperomia sp. 1	Herb	Native	Medicine
	Peperomia sp. 2	Epiphyte	Native	Charms
	Pipe anisatum HBK	Bush	Introduced	Medicine
	Piper sp.	Bush	Native	Medicine
	Piper tricuspe	Bush	Native	Medicine
	Pothomorphe peltata	Bush	Native	Medicine

(*continued*)

Table 6.2. Plant species grown in patios and zoteas (continued)

Family	Species	Habit	Origin	Use
	Pothomorphe umbellata	Bush	Native	Medicine
Plantaginaceae				
	Plantago major L.	Herb	Introduced	Medicine
Poaceae				
	Coix lacrimosa L.	Bush	Introduced	Crafts
	Cymbopogon citrates	Herb	Introduced	Food, medicine
	Saccharum officinarum	Bush	Introduced	Food, medicine
Polypodiaceae				
	Polypodiaceae	Herb	?	Soil cover
Portulacaceae				
	Portulaca oleracea L.	Herb	Native	Medicine
	Talinum sp. 1	Herb	Native	Food
	Talinum sp. 2	Herb	Native	Food
Rubiaceae				
	Borojoa patinoi Cuatr.	Tree	Native	Fruit, medicine
Rutaceae				
	Citrus limon Burm.	Tree	Introduced	Fruit, medicine, shade, fuel
	Citrus nobilis Loureiro	Tree	Introduced	Fruit, medicine, shade, fuel
	Citrus sinensis	Tree	Introduced	Fruit, medicine, shade, fuel
Sapindaceae				
	Melioccus bijugatus Jacq.	Tree	Native	Fruit, shade, fuel
Sapotaceae				
	Manilkara sp.	Tree	Native	Fruit, shade, fuel
	Pouteria caimito	Tree	Native	Fruit, shade, fuel

Table 6.2. Plant species grown in patios and zoteas (continued)

Family	Species	Habit	Origin	Use
Sellaginellaceae				
	Sellaginela sp.	Herb	Native	Charms
Solanaceae				
	Capsicum annuum L. var. 1	Bush	Native	Food, medicine
	Capsicum annuum L. var 2	Bush	Native	Food, medicine
	Cestrum nocturnum Lin.	Bush	Native	Ornamental
	Cestrum sp.	Bush	Native	Medicine
	Lycopersicon esculentum Miller	Bush	Introduced	Food, medicine
Sterculiaceae				
	Theobroma cacao	Tree	Native	Food, medicine, shade, fuel
Umbelliferace				
	Coriandrum sativum L.	Herb	Introduced	Food, medicine
Urticaceae				
	Urtica urens L.	Herb	Introduced	Medicine
Verbenaceae				
	Lantana camara L.	Bush	Native	Medicine
	Stachytarpheta cayanensis	Herb	Native	Medicine
	Verbena sp.	Herb	Native	Medicine
Violaceae				
	Viola sp.	Herb	Native	Ornamental, medicine
Zingiberaceae				
	Alpinia occidentalis	Herb	Introduced	Medicine
	Hedychium coronarium Koening	Herb	Introduced	Medicine, live fence
	Zingiber officinale Rose	Herb	Introduced	Food, medicine

Table 6.3. Animal species raised in patios and zoteas

Species	Use
Dogs	Hunting
Cats	Keeping mice away
Chickens (landraces and commercial races)	Eggs, meat
Ducks (commercial races)	Eggs, meat
Pigs (landraces)	Food, used as "piggy banks" for emergencies
Wild birds (parrots)	Luxury items
Wild animals (small mammals)	Luxury items

social organization and social networks. In the small villages of the Pacific, if a community member asks for a certain plant, it cannot be denied. Plants are always given as gifts; only a few herbs, such as green onions, are sold. A woman who does not grow plants or tend a garden is frowned upon because she cannot engage in reciprocal exchange relations. Disparagingly, an empty garden is likened to a naked woman.

Chocoan women's homegarden cultivation practices are based upon agronomic and cultural principles of soils, water, plant ecology, and the social function of plants. The spatial distribution of the plants around the house is based on plant use and habit. The front garden, which faces the principal communication and circulation avenue (the river or the street), is planted with bushy ornamental or "luxury" plants, as well as those used for protection against evil and for courting good luck. Luxury plants often lack local terminology because they are generally obtained from mestizos or brought from the interior of the country to emulate the Andean garden, well known for its exuberant ornamental vegetation. The space behind the house is the patio, where a random combination of herbaceous and brushy plants, as well as fruit and palm trees and a few crops, are planted, forming agroecosystems of varying complexities. Coconuts, papaya, lemon, pineapple, avocado, manioc, plantain, banana, *caimito* (*Pouteria caimito*), *borojo* (*Borojoa patinoi* Cuatr.), and *papachina* (*Colocasia esculenta* L., Schott) are some of the edible plants commonly found around the house. The patio is an open space surrounding the kitchen, where family members gather to talk while cooking or doing other domestic chores. In the patio are the elevated gardens, or zoteas, located close to the kitchen for convenience and security. Zoteas have two very important functions: they are used to cultivate medicinal, culinary, and aromatic herbs, and they also serve as nurseries to germinate fruit trees that are later transplanted to the agricultural fields, or *fincas*.

Raised gardens are evidence of local people's accumulated practical knowledge of and experience with various technologies for plant cultivation in tropical rainforests. They are also dynamic miniature agricultural domestic systems, which function as a link in the chain of energy flow that circulates through natural and cultural processes across the different spaces of the territory. Zoteas are made with old wooden canoes or wooden structures supported by wood from the forest or mangrove stilts. The plants and seeds are grown in a fertile substratum or compost made of decomposing leaves (*hojarasca*), rotting wood, sand, ashes, and ant soil from leaf-cutting ants found in the forest. The hojarasca is composed of forest leaves that fall in the river and are washed away into the ocean; during high tide they are deposited on the beach, where they are gathered together with sand and small pieces of driftwood. The elevated structure offers protection from domestic animals and floods and allows for drainage and air circulation. In areas with nutrient-poor soils, the use of fertile substratum is an efficient strategy that takes advantage of local organic resources for cultivation. The substratum is a pH-neutral soil rich in potassium and phosphorus, as well as plant-fungi interactions (mycorrhizae), which enhance nutrient assimilation and plant growth. The germination of seeds under these conditions gives a comparative advantage to the plants and fruit trees that are later transplanted to forest areas with low-fertility conditions and compact soils. The soils and the surrounding wild vegetation in different habitats are in turn improved by the presence of the mycorrhizae present in the seedlings.

In a hardly noticeable but very efficient manner, homegardening knowledge and experimentation connect the domestic space with the forest and contribute to the enhancement of biodiversity throughout the territory, as more resilient plants are propagated. Fruit tree planting also reaffirms the cultural appropriation of the territory by different means: unlike plants that self-propagate, such as bananas and plantains, or "walking trees," fruit trees with steady roots have been used as boundary markers between family fields, local Afro-Colombian communities, and indigenous lands. Because trees are individually owned and inherited, they are an important part of local customary rights, upon which black communities have based their legal land claims. Homegarden management is thus a vital aspect of the spatialization of social and environmental relations in domestic settings and beyond. Women's plant cultivation practices play a significant role in the context of territorial appropriation and sociocultural reconstruction in the Pacific Chocó coast.

Contrary to outside notions of the Pacific as a vast empty forest reserve, it is an occupied and meaningful place, marked by people's everyday interactions with each other and with nature. Recently, however, Afro-Colombian homeplaces and communities have been disrupted by the forced internal displacements occurring in the Pacific as the result of armed disputes over the control of this geographically strategic and resource-rich area by various actors with distinct economic and political interests. This displacement has been termed a "humanitarian catastrophe" (Rosero 2002) because of the dramatic socioeconomic and cultural consequences for the vulnerable ethnic populations caught in the middle of a war in which they do not belong, as this testimony from a displaced Afro-Colombian woman poignantly demonstrates:

> Violence against women is to take away their right to their territory, is to force women to leave their territory, attempt against us, our culture, our territory, our way of life, force us to submit to different rhythms to those of our culture. It is this discrimination to which we have been subject since times past, also the racial discrimination which society forces on us, is the mistreatment we receive for being displaced and being black. (Instituto de Servicios Legales Alternativos 2002:15)

Making Homeplace at the Margins

Against the backdrop of displacement, Afro-Colombian women have claimed and marked their places, through whatever means available to them. bell hooks (1990) has argued that homeplace, however fragile, has never been politically neutral for Afro-descendents because homes were places that enabled slaves to become dignified subjects, rather than objects, and where people could return to heal and recover their wholeness and humanity. More broadly, homeplaces facilitated the constitution of communities of resistance and solidarity in the midst of a situation of exclusion and domination. In the context of heightened uncertainty, as was the case of Africans in the Americas in the sixteenth and seventeenth centuries, and the current situation of rural Afro-Colombians, the construction of a secure and safe place is a fundamental part of an ongoing political power struggle. As hooks notes, homeplace making was a central dimension of black women's everyday domestic labor in the United States and had a powerful political meaning vis-à-vis the hostility of the public world.

In Colombia, black women's history is inscribed in a context of patriarchal power, colonial domination, cultural and spatial fragmentation, and structural marginalization. Since their inception as slaves valued for the productive function of their bodies and the reproduction of free labor for slave owners, black women have been central to the reconstitution of the black family and kinship networks, the recreation of the spiritual and cultural world, and in "re-membering" the social fabric through home-place making.[4] Although Afro-Colombian social organization and family structure is the result of specific regional developments, women have had a salient role since colonial times in consolidating descent and domestic systems. Scholars explain matrilineal kinship patterns and a positive valuation of a prolific maternity and extended family as the result of the identification of children of slave women through maternal lines, because women were forced to have multiple sexual partners in slave camps and paternal relations were not encouraged by slave owners (Gutiérrez 1968; Motta 1993; Romero 1995; Zuluaga and Bermúdez 1997). Mothers and grandmothers were the most stable and cohesive social figures, and their multiple unions became a strategy to expand alliances through feminine lines. The high spatial mobility of black men, who historically engaged in various productive activities (mining, fishing, hunting, off-farm wage labor) throughout the year, contributed to the reproduction of this particular form of social organization. While young Afro-Colombian women were also highly mobile, mature women tended to stay in place, where their domestic and productive roles were visible and socially recognized. Much of their activism focused on achieving and maintaining rights to and control over spatial resources that sustained everyday life and livelihoods.

During slavery, homeplace making in the Colombian Pacific took different forms according to local environmental and socioeconomic conditions. A shared element was the constitution of domestic communities in slave camps and *palenques*, or refuge sites, established by runaways deep into the forest, which operated as communities of resistance with varying degrees of success. These were occasionally shared with indigenous groups with whom slaves shared escape routes (Vargas 1993). Upon manumission, the home-making process continued with the rapid colonization of marginal and uninhabited or sparsely populated areas along rivers and estuaries. In the Chocó, the dominant settlement pattern was the establishment of indigenous villages in inaccessible riverheads on the mountains and black villages along rivers and coasts. Afro-Colombians developed subsistence economies based on a multiple, flexible, and seasonal exploitation of different ecosystems and resources. The landscape was appropriated following a clear division of labor: first men cleared or "tamed" the wild primary forest (*monte bravo*), and then women joined in the planting of staples such as corn, plantain, various tubers, and fruit trees. Traditionally, hunting, fishing, and clearing the forest have been masculine activities, while women are in charge of creating and tending the domestic spaces of the home and surrounding homegardens.

The physical appropriation and transformation of the landscape were matched by the symbolic appropriation of space through place naming.

Mountains, valleys, river courses, beaches, and settlements were endowed with a rich and sonorous toponymy of mixed indigenous-, Spanish-, and African-derived words. Over time the landscape has been transformed into a readable entity embodying people's particular histories, social relations, moral teachings, and ecological knowledge. The relationship of Afro-Colombian men and women with the landscape is reminiscent of the situated intimacy of Amazonian river dwellers described by Hugh Raffles (2002), which involves local knowledge of nature rooted in practice, experience, and, most important, affect.

In 1952, the national government designated these marginal areas as empty national lands (*baldíos nacionales*) to be protected under the national forest reserve legislation. Since the term *baldío* literally means empty, this measure explicitly overlooked the existence of black and indigenous communities and their rights to the territories they occupied. By assuming sovereign control of these strategic areas and the resources therein, the state claimed the exclusive right to use and/or grant concessions or property titles to third parties. This legislative measure was disputed in the late 1980s when local communities and organizations initiated their territorial claims on the grounds of an ancestral occupation of the land and the cultural practices that have enabled the perpetuation of peoples and ecosystems through time.

The *Ombligada*: Embodying and Placing Gender Identity

Throughout the material process of place making and the symbolic and ritual appropriation of the territory and its resources, Afro-Colombians have developed an extensive environmental knowledge and have carved out their individual and collective ethnic and gender identities. Black women play a significant role in life cycle rituals, which reflect a group's particular cosmology and serve to establish a specific relationship with the world and society. At birth midwives and grandmothers in the Pacific perform initiation rituals to welcome the newborn, and upon death they sing ritual songs (*alabaos*) and prayers during the wake to facilitate the transit of the dead into the afterworld. A common practice when a child is born is to bury the placenta and the umbilical cord under a tree to connect the newborn to a specific place. This first literal and symbolic grounding provides a sense of belonging, and people often identify home as the place where the navel is buried.

Another salient rite of passage in the Pacific known as the *ombligada*, or navel curing, is performed on the newborn soon after. In this ritual, elements from the environment are incorporated materially and symbolically into a person's body to endow him/her with the attributes and powers of each substance. This propitiatory rite is carried out by mothers, grandmothers, or midwives after a child's birth and consists of placing a series of ground animal, vegetable, or mineral substances into the infant's navel. The navel curing is done several times a day until the navel is dry. The body of the child becomes a receptacle for the powers of the local environment and the embodied powers are later expressed in the form of personality traits, behaviors, or attributes for succeeding in life. For instance, tapir's claw provides strength, tree bark provides endurance, and earthworms or eels grant protection during fights by making the person's body "slippery." Some of these items are used for both girls and boys, for whom physical strength is desirable. Men and women experience great pride and pleasure with a strong, healthy, and well-shaped body. Hard labor, economic independence, and spatial mobility are also positively valued among men and women as qualities that will facilitate their survival and success during times of hardship. However, certain elements, like silver or gold, which may work positively to bring luck and fortune to men, may cause girls to never settle down but always be on the move, without any stability, and running from the hands of men like money does. Following the same logic, some substances are specifically employed to endow girls with desired feminine qualities and skills or propitiate positive outcomes. This includes items such as wood from a kitchen spoon to be a good cook, wood from a house post to have a home in adulthood, or a medicinal plant to have curing powers. These products are also found in domestic or feminine spaces. Some of the substances used for boys are only found in the areas where men's activities take place: the sea and the forest. Places, the resources associated with them, and the meanings of those places are thus embodied in the individual.

As Arturo Escobar has noted with respect to the articulation of place, body, and environment, "all cultural practices are emplaced and culture is carried into places by bodies" (2001:143). The ombligada is another way to ground people physically and symbolically in place while reaffirming the gendered nature of space and reasserting particular gender identities through the use of specific products from the surrounding environment (Losonczy 1989). The burial of the umbilical cord and the placenta and the curing of the navel connect people with their natural surroundings in concrete embodied and emplaced cultural practices. Nurturing and caring feminine practices like these among blacks in the Colombian Pacific, which foster familiarity and intimacy with the surrounding environment, are reminiscent of hooks's argument about the role of black women in the creation of a much-needed sense of belonging, grounding, and identity for marginalized African Americans.

Displacement and the Challenge of Reconfiguration

Henri Lefebvre has noted that "socio-political contradictions are realized spatially [and] spatial contradictions 'express' conflicts between socio-political interests and forces" (1991:365). In the Colombian Pacific, paradoxically, just as the long process of homeplace making began to be formalized with the recognition of Afro-Colombian ethnoterritorial rights, local populations began to experience the impact of a brutal war that has forced them to abandon their homelands. Displacement is part of an economic and political strategy on the part of various interest groups that dispute territorial control over the geopolitically strategic and resource-rich areas (including minerals, timber, biodiversity, marine resources) and the land available for large development projects and agribusiness (African oil palm and coca). For Afro-Colombians, the current forced internal displacement is part of a series of systematic and deliberate actions conducive to the war and to the development model, which sees in the forest a vast empty space to be transformed for energy, mining, communication, and recreational enterprises (Rosero 2002). In the last few years this dispute in the Pacific region has not only grown but also come to involve flagrant violations of human rights by means of intimidation and terror that include homicide, rape, and forced displacement (Instituto de Servicios Legales Alternativos 2001; Encuesta Nacional de Verificación de Población Desplazada 2008).

Some of these violations were long anticipated, but official pre- and post-conflict responses to displacement have been insufficient and dispersed. Those who decide to resist in place or to return face restricted mobility, controlled access to food, forced labor and sexual services, and the constant threat of death. Those who leave find themselves in a state of great economic, nutritional, health, psychological, and social vulnerability (Rapoport Center 2007; Encuesta Nacional de Verificación de Población Desplazada 2008). By disrupting and uprooting communities and social networks, displacement creates new notions of home and identity. Women's frustration and aspirations are summarized in a succinct testimony: "We want to live free. Armed actors force us to live without feet to walk, without hands to work, and without mouths to talk" (Instituto de Servicios Legales Alternativos 2002:17).

According to government estimates, the displaced population can reach up to 4 million people. Despite the lack of complete and reliable statistics, and disaggregation by ethnicity, there is consensus that both indigenous and black communities from the Pacific are among the most affected communities by this chronic crisis. Official reports (Encuesta Nacional de Verificación de Población Desplazada 2008) estimate that one-fourth of the total displaced population is of Afro-Colombian origin and 98 percent of this group falls below the poverty line. Half of those displaced are women; many are also heads of households. For ethnic groups, forced displacement means breaking away from the collectivity and the territory, from the land that provides the means of subsistence and the sense of identity and belonging. Ethnic organizations and social movements are also experiencing political and organizational dispersal. Threats and murders of national and local leaders have weakened their ability and strength to mobilize politically for what is known as "the ancestral mandate": defense of the territory, identity, and autonomy.

With the dismemberment of the community and the unavailability of refuge areas there is a loss of the material and moral conditions for the development of personal and collective life projects. The options for cultural reconstitution and territorial thought are minimal to no longer existent. This devastating situation means the loss of those sites of resistance, dignity, and solidarity women have helped to build with their work, their songs, and their memories. The current destruction of physical and symbolic homeplace and social and territorial autonomy has been identified by black leaders as one of the greatest aggressions against Afro-descendents in the past 150 years (Rosero 2002). Afro-Colombian men have been systematic targets of violence by armed groups. Women and girls are also subject to physical and psychological violence; sexual violence is one of the most common and largely unreported crimes. For women, the war dynamics reinforce a situation of structural exclusion and discrimination on the basis of race, gender, and the new *desplazado* (displaced) condition, which is ambiguously perceived as a victim of violence but at the same time a potential threat. Family migration often increases intrahousehold violence (Instituto de Servicios Legales Alternativos 2001). Women often are forced into the informal sector, as domestic servants or prostitutes, to support their families. As desplazadas put it, to be displaced is to be at the mercy of others (Afrodes—Global Rights 2008).

Displaced Afro-Colombians are faced with the challenge of reconfiguring fragmented collectivities and restituting their wholeness and dignity in a new process of homeplace making at the margins. In this situation women must once again reconstitute the black family and kinship networks, recreate the spiritual and cultural world, and re-member the social fabric through homeplace making. They must craft new social identities, relations, and skills to navigate and survive in unfamiliar hostile settings, which are primarily urban and mestizo. The loss of familiar referents necessitates the reorganization of social and spatial relations in *nonplaces*, defined by Linda McDowell as "those locations in the contemporary world where the transactions and interactions that take place are between anonymous individuals often stripped of all symbols of social identity" (1999:6). Black women often find that the gendered knowledge and practices that sustained them in rural contexts are inappropriate and limit the possibilities of urban economic and social reestablishment. Without clear material and symbolic referents and a sense of trust in anything and anyone, often women find themselves lacking direction.

Displaced black women are not passive or helpless victims. At a political level, they have organized to agitate for the recognition, restitution, and reparation of their fundamental human rights within a gender perspective (Afrodes—Global Rights 2008). In the daily struggle for inclusion and exclusion in new urban environments, they have also resorted to a creative use of knowledge, strengths, and skills to maximize material and symbolic resources to appropriate and resignify new spaces. In their attempts to carve out new homeplaces and re-member fractured subjectivities, new social and environmental relations are spatialized in domestic and public ambits, and novel cultural forms and identities emerge. In the midst of tensions and contradictions, women capitalize on ethnic stereotypes of black women's domesticity to find economic opportunities as maids, cooks, street food vendors, and small merchants, thereby establishing family and regional networks and "ethnicizing" the informal economy, the labor market, the cultural composition of cities and neighborhoods, and the tastes and culinary choices of urban communities.

Afro-Colombian women's agency and their practices of everyday resistance continue to focus on the material, spatial, and symbolic resources of their (new) homeplace, in both the public and private domain. Their strength is sustained by a tradition of struggle for the right To Be in place in pervasive contexts of social inequities and racial discrimination. The question, which bears further investigation, is to what extent their homeplace making will carry them through these trying times and if it does, what lessons such strategies can teach the rest of humanity, which is itself displaced and dislocated in various ways. Understanding women's use of space and ways of emplacement—their homegardens, zoteas, and symbolic rituals—can provide us valuable insights on reconfiguring identities under siege.

Notes

1. Afro-Colombians are a very diverse population, and so are their historical, environmental, socioeconomic, and cultural experiences. I draw primarily on data from fishing and agricultural communities in the northern Pacific coast.

2. Mestizos or "mixed blood" are people of mixed indigenous and white descent and today constitute the majority of the Colombian population.

3. Taming does not mean violent confrontation, domination, or domestication but transformation. Afro-Colombians do not see their relationship with nature

as one of human control over it. By this, however, I do not want to romanticize their environmental relationships or characterize all their environmental actions as benign or sustainable.

4. Elsewhere I have argued that the Colombian literature on Afro-Colombian women, mostly written by mestizos and according to white/mestizo notions of femininity, is still incipient, fragmentary, and riddled with contradictory and essentialist representations that range from the mythical matriarch to the victimized slave (Camacho 2004).

References

Afrodes—Global Rights. 2008. *Vidas ante la adversidad. Informe sobre la situación de los derechos humanos de las mujeres afrocolombianas en situación de desplazamiento forzado.* Bogotá: Afrodes—Global Rights.

Aguilar-Stoen, Mariel, Stein R. Moe, and Sara Lucía Camargo-Ricalde. 2009. Homegardens sustain crop diversity and improve farm resilience in Candelaria Loxicha, Oaxaca, Mexico. *Human Ecology* 37(1):55–77.

Alberico, Michael. 1993. La zoografia terrestre. In *Colombia Pacífico*, edited by Pablo Leyva, 1:232–238. Bogota: Fondo Financiera Energética Nacional.

Asher, Kiran. 2009. *Black and Green: Afro-Colombians, Development, and Nature in the Pacific Lowlands.* Durham, NC: Duke University Press.

Camacho, Juana. 1999. Todos tenemos derecho a su parte: Derechos de herencia, acceso y control de bienes en comunidades negras de la costa Pacífica chocoana. In *De montes, ríos y ciudades: Territorios e identidades de la gente negra en Colombia*, edited by Juana Camacho and Eduardo Restrepo, 107–130. Fundación Natura-Ecofondo-Instituto Colombiano de Antropología. Bogotá: Giro Editores.

———. 2000. Mujeres, zoteas y hormigas arrieras: Prácticas de manejo de flora en huertos de la costa chocoana. In *Zoteas: Diversidad y relaciones culturales en el Chocó Biogeográfico Colombiano*, edited by Jesús E. Arroyo and Juana Camacho. Instituto de Investigaciones Ambientales del Pacífico-Fundación Natura-Swissaid. Medellín: Intempo.

———. 2004. Silencios elocuentes, voces emergentes: Reseña bibliográfica de los estudios sobre la mujer afrocolombiana. In *Panorámica afrocolombiana. Estudios sociales en el Pacífico*, edited by Claudia Mosquera, M. Clemencia Ramírez, and Mauricio Pardo, 167–210. Bogotá: Instituto Colombiano de Antropología e Historia—Universidad Nacional de Colombia.

Camacho, Juana, and Carlos Tapia. 1997. Black women and biodiversity. Report submitted to the MacArthur Foundation. Bogotá: Fundación Natura.

Domosh, Mona. 1998. Geography and gender: Home again? *Progress in Human Geography* 22(2):276–282.

Encuesta Nacional de Verificación de Población Desplazada. 2008. Proceso nacional de verificación de los derechos de la población desplazada. Primer informe a la Corte Constitucional. Comisión de Seguimiento a la Política Pública sobre el desplazamiento forzado. Bogotá: Comisión de Seguimiento— Consultoría para los Derechos Humanos y el Desplazamiento—Centro de Investigaciones para el Desarrollo.

Escobar, Arturo. 2001. Culture sits in places: Reflections on globalism and subaltern strategies of localization. *Political Geography* 20:139–174.

———. 2008. *Territories of Difference: Place, Movements, Life, Redes.* Durham, NC: Duke University Press.

Escobar, Arturo, and Alvaro Pedrosa. 1996. *Pacífico Desarrollo o diversidad?* Bogotá: Ecofondo-Cerec.

Escobar, Arturo, Dianne Rocheleau, and Smitu Khotari. 2002. Environmental social movements and the politics of place. In *Place, Politics and Justice: Women Negotiating Globalization*, edited by Wendy Harcourt, 28–36. Society for International Development 45. London: Sage.

Eyzaguirre, Pablo, and Olga F. Linares, eds. 2004. *Home Gardens and Agrobiodiversity.* Washington, DC: Smithsonian Books.

Feld, Steven and Keith H. Basso. 1996. *Senses of Place*, Advanced Seminar Series. New Mexico: School of American Research Press.

Flores, Jesus A., and Constanza Millan. 2007. *Derecho a la alimentación y al territorio en el Pacífico colombiano.* Bogotá: Diócesis de Tumaco, Quibdó, Buenaventura e Istmina.

Gentry, Alwin. 1993. Riqueza de especies y composición florística. In *Colombia Pacífico*, edited by Pablo Leyva, 1:200–219. Bogotá: Fondo Financiera Energética Nacional.

Grueso, Libia, and Leyla A. Arroyo. 2002. Women and the defence of place in Colombian black movement struggles. In *Place, Politics and Justice: Women Negotiating Globalization*, edited by Wendy Harcourt, 60–67. Society for International Development 45. London: Sage.

Gupta, Akhil, and James Ferguson, eds. 1997. *Culture, Power, Place. Explorations in Critical Anthropology.* Durham, NC: Duke University Press.

Gutiérrez de Pineda, Virginia. 1968. *Familia y cultura en Colombia. Tipología, funciones y dinámica de la familia. Manifestaciones múltiples a través del mosaico cultural y sus estructuras sociales.* Bogotá: Tercer Mundo Editores.

hooks, bell. 1990. Homeplace: A site of resistance. In *Yearning: Race, Gender, and Cultural Politics*, edited by bell hooks, 41–49. Boston: South End Press.

Howard, Patricia, ed. 2003. *Women and Plants: Gender Relations in Biodiversity Management and Conservation.* London: Zed Press and Palgrave-Macmillan.

Ingold, Tim. 2000. *The Perception of the Environment. Essays on Livelihood, Dwelling and Skill.* London: Routledge.

Instituto de Servicios Legales Alternativos. 2001. *Informe sobre violencia sociopolítica contra mujeres y niñas en Colombia Segundo avance-2001.* Mesa de Trabajo Mujer y Conflicto Armado. Bogotá: Antropos.

————. 2002. Efectos del conflicto armado en las mujeres afrocolombianas. Draft report. Bogotá: Instituto de Servicios Legales Alternativos.

Jimeno, Gladys. 2001. Exodo e identidad. In *Exodo, Patrimonio e identidad. Memorias de la V Cátedra Anual de Historia Ernesto Restrepo Tirado*, 422–436. Bogotá: Ministerio de Cultura.

Lefebvre, Henri. 1991. *Critique of Everyday Life*. London: Verso.

Losonczy, Anne Marie. 1989. Del ombligo a la comunidad: Ritos de nacimiento en la cultura negra del litoral Pacífico colombiano. *Revindi* 1:49–54.

McDowell, Linda. 1999. *Gender, Identity and Place: Understanding Feminist Geographies*. Minneapolis: University of Minnesota Press.

Motta, Nancy. 1993. Mujer y familia en la estructura social del Pacífico. *Revista Colombiana de Trabajo Social* 6.

Nazarea, Virginia D. 2005. *Heirloom Seeds and Their Keepers: Marginality and Memory in the Conservation of Biological Diversity*. Tucson: University of Arizona Press.

Oakley, Emily, and Janet Momsen. 2007. Women and seed management: A study of two villages in Bangladesh. *Singapore Journal of Tropical Geography* 28(1):90–106.

Palsson, Gisli. 1994. Enskillment at sea. *Man* 29:901–927.

Pollock, Griselda. 1988. *Vision and Difference: Femininity, Feminism and the Histories of Art*. London: Routledge.

Proceso de Comunidades Negras. 1996–1998. El concepto de territorio en las comunidades negras del Pacífico centro y sur. Unpublished document. Bogotá.

Raffles, Hugh. 2002. Intimate knowledge. *International Social Science Journal* 173:325–334.

Rapoport Center. 2007. Unfulfilled promises and persistent obstacles to the realization of the rights of Afro-Colombians. A report on the development of Ley 70 of 1993 submitted to the Inter-American Commission on Human Rights. University of Austin, Texas.

Restrepo, Eduardo. 1996. Tuqueros negros del Pacífico sur colombiano. In *Renacientes del Guandal: "Grupos negros" de los ríos Satinga y Sanquianga. Pacífico sur colombiano*, edited by Del Valle and E. Restrepo, 1:243–348. Bogotá: Biopacífico-Universidad Nacional.

————. 2001. Imaginando comunidad negra: Etnografía de la etnización de las poblaciones negras en el Pacífico sur colombiano. In *Acción colectiva, estado y etnicidad en el Pacífico colombiano*, edited by Mauricio Pardo. 41–70. Bogotá: Icanh-Colciencias.

Rocheleau, Dianne, Barbara Thomas-Slayter, and Ester Wangari, eds. 1997. *Feminist Political Ecology: Global Perspectives and Local Experience*. London: Routledge.

Romero, Mario Diego. 1995. *Poblamiento y sociedad en la costa Pacífica colombiana siglos 17–18*. Cali: Departamento de Historia, Universidad del Valle.

Rose, Gillian. 1993. *Feminism and Geography: The limits of geographical knowledge*. Minneapolis: University of Minnesota Press.

Rosero, Carlos. 2002. Los afrodescendientes y el conflicto armado en Colombia: La insistencia en lo propio como alternativa. In *Afrodescendientes en las Américas: Trayectorias sociales e identitarias—150 años de la abolición de la esclavitud en Colombia*, edited by Claudia Mosquera, Mauricio Pardo, and Odile Hoffmann, 547–559. Bogotá: Universidad Nacional.

Soja, Edward W. 1980. The socio-spatial dialectic. *Annals of the Association of American Geographers* 70: 207–225.

Vargas, Patricia. 1993. *Los embera y los cuna impacto y reacción ante la ocupación española siglos XVI y XVII*. Bogotá: Instituto Colombiano de Antropología— Centro de Estudios Regionales Colombianos.

Wouters, Mieke. 2001. Ethnic rights under threat: The black peasant movement against armed groups' pressure in the Chocó, Colombia. *Bulletin of Latin American Research* 20(4):498–519.

Zuluaga, Francisco, and Amparo Bermúdez. 1997. *La protesta social en el suroccidente colombiano: Siglo 18*. Cali: Instituto de Altos Estudios Jurídicos y de Relaciones Internacionales, Universidad del Valle.

PART TWO

Agency and Reterritorialization in the Context of Globalization

CHAPTER SEVEN

Maya Mother Seeds in Resistance of Highland Chiapas in Defense of Native Corn

PETER BROWN

With careful examination of the tiny test strip, everyone present convinced themselves that a pink line indicating contamination by genetically engineered corn was missing. As this individual corn leaf tested free of contamination from genetically modified organisms (GMOs), a palpable sigh of relief seemed to fill the room and flow through the crowd of tense Mayan men and women. The test strip was carefully affixed to the previously labeled plastic bag of chemical solvent and newly picked corn leaves while the assembly examined another test strip. The long day of discussion and corn testing was an emotional roller coaster for everyone as several dozen farmers from the two ethnic groups of the northern region of Chiapas tabulated results and picked the brains of the small team of nonindigenous visitors cooperating with the Mother Seeds in Resistance project.

"We are happy when our corn is still natural," explained one young indigenous woman who had spent that entire spring day exploring the intricacies of field testing corn for transgenic markers. "There is still much that we need to learn about these genetic modifications, but we know we do not want what the big corporations are sending because we want our food to be pure and our corn to be natural."

Our small delegation of Mexican and U.S. activists had arrived several days earlier to attend the dedication of a major new education center the Zapatistas had constructed over the last three years using funds raised by our parent organization, Schools for Chiapas. In addition, we carried with us several dozen kits for the field testing of corn for transgenic contamination, along with a heavy cargo of instructions, suggestions, and information from our departing staff scientist, Martin Taylor. This is a deeply personal and emotional tale rather than a detailed scientific account, inspired by the words of a Zapatista education promoter at the National Forum in Defense of Mexican Corn in Mexico City: "For us, the indigenous, corn is sacred. If these agrochemical companies are trying to get rid of our corn, it is like wanting to get rid of a part of our culture which we inherited from our Mayan ancestors. We know that corn is our primary and daily food, it is the base of our culture" (January 2002).

Staff Scientist Is Lost to Love

For two years Martin Taylor had poured his extraordinarily focused energy and internationally acclaimed scientific expertise into Mother Seeds in Resistance from the Lands of Chiapas. On the heels of winning his university tenure battle, Martin drove our ancient pickup truck from his home in Tucson to Chiapas. Once there, he endured numbingly long bus rides, taught innumerable classes to indigenous teachers and students about transgenic contamination, testified before NAFTA-created environmental bodies, dreamed up innovative laboratory techniques for long-term preservation of corn, and designed extensive field-testing protocols appropriate for indigenous groups in Chiapas. It seemed that nothing could derail Martin until a visit from Cupid sent him racing off to his ancestral home in Australia with new bride in tow, his head filled with visions of hearth, home, and family.

Like most Schools for Chiapas personnel, Martin was a volunteer. In fact, in an act of extraordinary generosity, he used his own money to purchase project equipment such as the freezer that now stores indigenous seeds in long-term storage in the highlands of Chiapas. Like our hosts, the indigenous Zapatistas of Chiapas, Schools for Chiapas has always been more successful in generating visions and hope than in raising funds. Naturally, there was no budget to hire a new staff scientist when Martin succumbed to Cupid's call.

After all of Martin's careful instruction and encouragement, our team felt capable of carrying on the work. Still, his absence left our team of nonscientists a bit nervous as the Good Government Board of the north—whose name, "The New Seed Which Is Going to Produce," is always publicly presented in three languages of Chiapas—Spanish (Nueva Semilla que Va a Producir), Tzotzil (Yach'il Ts'unubil Yuun Yax P'plj), and Chol (Tsi Jiba Pakabal Micajel Polel)—began to contact communities by radio and individual messengers urging an immediate collection of seed for national and international solidarity distribution, as well as collection of corn leaves so that testing for GMO contamination could proceed in the north of Chiapas.

"Here the People Command and the Government Obeys"

It was with some trepidation on that morning in April 2004 that I watched as Chol and Tzeltal farmers arrived in droves for the religious and civil school dedication ceremonies that were to take place in the government center. During massive celebrations just eight months earlier, the autonomous, indigenous government of north Chiapas had assumed in three languages their provocative yet fitting name. Now it seemed appropriate that the Juntas de Buen Gobierno (Good Government Board) serving the entire north of Chiapas would call upon their communities to bring a part of their first corn harvest to this important community gathering. However, the celebratory school dedications and the productive spring harvest were overshadowed by threats of GMO contamination—a threat imported from my country.

Throughout the morning, dozens of farmers and their families arrived carrying the fruits of months of backbreaking agricultural labor. As the burning tropical sun rapidly gained strength, I was struck by the similarity in the almost ceremonial entrance of every farming family that arrived from several counties governed by the insurrectionary Zapatista governments. No sooner would a dusty pickup crammed full of indigenous people arrive than several men would leap out and then reverently turn back to receive packages of corn leaves and bags of multicolored seed from the hands of sons and daughters, mothers and grandparents.

The unique mixture of joy and sadness that had marked my mood all morning deepened with each arrival. Everyone shared the excitement about the colorful seed corn designated for solidarity encompassing supporters across Mexico and around the world. Everyone realized that extensive contaminations might be discovered during the planned genetic testing and that these distant "grow-outs" were necessitated by the threat of genetic contamination.

Although each genetic test required only a small leaf sample, occasionally a farmer who had not fully understood instructions would bring an entire ten-foot-high corn plant uprooted from his jungle *milpa*, or field, and adorned by several massive corn ears. After several of these beautiful mature plants ended up resting against the mural decorating the offices of the municipal governments, my spirits rose dramatically. As bursts of howler monkey cries endorsed her flight, a joyful white peace dove at the center of the mural appeared to be propelled into flight by the shining stalks of powerful Mayan corn. I settled down to participate in the greeting of each farmer inside the municipal building, which doubled as our dormitory and genetic testing laboratory.

Inside this tin-roofed and dirt-floored wooden building, the ceremony of corn continued and deepened. After receiving their corn packages from the arriving trucks, each man ambled toward the colorful but ramshackle office of their county governments. They entered silently, without bothering to knock or ask permission. After their eyes adjusted to the relative darkness of the windowless building, the newcomers silently surveyed the scene and greetings were exchanged in Chol or Tzeltal. Occasionally a Spanish word or two crept into these initial greetings when the official and the newly arriving farmer happened to not speak the same language.

For our team of outsiders the silences between greetings often seemed uncomfortably long, the wait for an official of the appropriate language group interminable, and the ensuing exchange in incomprehensible languages, well, incomprehensible. But indigenous Chiapas is a time machine where life is viewed in generations and a calm dignity seems to extend tolerance to infinity. With the same indigenous dignity, detailed explanations were offered by Zapatista officials and carefully considered by Zapatista men and women.

Zapatista officials explained their request that farmers donate seed for grow-out and safekeeping, their project to test corn for transgenic contamination, and their decision to send seed outside of Chiapas. After actively participating in this process with questions and comments, the men often returned with wives, sons, daughters, and grandparents, initiating a new round of greetings, explanations, questions, and commentary. Although I could understand few of the Chol and Tzeltal words, the dignity and respect evident in the exchanges between officials and visiting farmers made me think of the proclamation that graces the ubiquitous Zapatista signs along roads and beside towns throughout Chiapas. From the highlands to the jungles, these signs of the autonomous indigenous Good Government Boards always declare, Esta Usted en Territorio Zapatista en Rebeldia—Aqui Manda el Pueblo y el Gobierno Obdece (You Are in Zapatista Territory in Rebellion—Here the People Command and the Government Obeys).

Zapatista Adventures in Genetic Testing—Part 1

Later in the day when our team was told that the visiting farmers were ready to meet, we hurried to gather our entire team in the county government headquarters. The Schools for Chiapas team arrived about the same time as dozens of indigenous men and women. Immediately our small and nervous group of outsiders received an enthusiastic welcome in Spanish from representatives of the Good Government Board. Everyone paid close attention as the warm official greeting was carefully translated into Chol and Tzeltal.

Then the autonomous authorities turned toward the several dozen assembled community representatives and began: "We the autonomous indigenous authorities of 'The New Seed That Is Going to Produce' have asked you to bring your corn seed and your corn leaves today to begin a new resistance to the transgenic contaminations that are being sent here from the United States by the big companies with the support of the government."[1] The night before, a Zapatista official had told our group of his personal concerns for the purity of his corn crop and the health of his community:

> You have heard the welcome of the Good Government Board to those from Mother Seeds in Resistance who have traveled here from far away to teach us about transgenic corn. Our friends tell us that some people in Mexico and some people in other parts of the world will grow our corn in their own lands. We believe this is a good thing. We also think that it will be good for us today to begin to learn how to test our corn to see if it is pure. Without these tests we cannot see the transgenics, which are attacking our corn. Thank you for listening. I hope that you have understood what I mean to say and that nothing I have said has offended anyone. Thank you, and that is all my words.

I felt slightly dizzy as I saw another Zapatista adventure beginning to unfold before my eyes. I was certain this new adventure would offer the profound challenges and rewards that members of Schools for Chiapas have repeatedly experienced over ten years of working with the autonomous Mayan peoples of Chiapas. And I was certain that all of us would be changed in the process.

Background to the Zapatista Movement

The civilian center where the events described above unfolded is one of five Zapatista centers in Chiapas known as Caracoles, which was previously known as Aguascalientes. For Zapatistas, the *Caracol* or "snail" designates a place where the inside meets the outside—where Mayan communities can come to know indigenous and nonindigenous communities from throughout the world. In addition to serving as meeting places, the Caracoles are home to Indian-run hospitals and schools, indigenous government offices and independent peace observers, Maya projects such as women's cooperatives and coffee producers' cooperatives, and myriad basketball courts and various sports fields.

The Zapatista movement burst on the international and Mexican scene on New Year's Eve 1994, with the almost bloodless military occupation of the five major cities of the highlands of Chiapas. Thousands of poorly armed, mostly monolingual, indigenous rebel soldiers poured into the cities as tens of thousands of unarmed Mayan community members felled trees and dug ditches across major highways to slow the inevitable advance of Mexican government troops. The rebels immediately proclaimed a law guaranteeing rights for indigenous women and denounced the North American Free Trade Agreement (NAFTA), which had taken effect that very morning. Specifically, the Zapatistas insisted that if fully implemented, NAFTA would mean the end of indigenous life in Chiapas and throughout Mexico.

The Mayan rebels who occupied cities certainly engaged the undivided attention of the national and international media, and of the Mexican military. Headlines and colorful photos throughout the world described the armed insurrection as the indigenous response to the first day of NAFTA while tens of thousands of unarmed Mayan peasants, Zapatista and non-Zapatista alike, moved to seize and occupy farming lands through out the southern most Mexican state of Chiapas. As the armed Mayan rebels began an orderly retreat from the five cities, many retreated to land seized by their kith and kin during the occupation of the cities and the battles that followed. As the late political journalist and Chiapas lawyer Amado Avendaño was often heard to publicly comment, "In that one short evening the Zapatistas gained more justice for the indigenous of Chiapas than I have been able to win in thirty years representing indigenous clients before Mexican courts."

Meanwhile, the words of Zapatista spokespeople and their documents flew around the world on the electronic wings of the Internet, combating the misinformation spread by the Mexican establishment and the CIA about the rebels. After just two weeks of heavy fighting in Chiapas, the Mexican people rejected the arguments of their government and demanded peace in the streets of every major city of the nation. The Mexican legislature passed the Law of Peace and Dialogue, which recognized the Zapatistas as a legitimate social organization, obliged Mexican armed forces to discontinue offensive actions, and mandated that the government open a peace dialogue with the rebel Indian forces.

Since that time, the Zapatistas have honored the cease fire and searched for nonmilitary solutions to the conflict while steadfastly maintaining their demands for justice, dignity, and democracy. The Zapatistas and representatives of the Mexican government have signed peace accords only to have that agreement abrogated by the Mexican government. Yet the Zapatistas have continued to honor the cease fire despite well-documented attacks by government-backed forces, which have resulted in more than 30,000 displaced Zapatistas who have been forced into hiding and refugee camps.

The Zapatistas have undertaken a number of innovative political initiatives, such as the 1999 independent plebiscite in which 2,500 Zapatista men and 2,500 Zapatista women, one man and one woman traveling together, visited every county in Mexico to argue for their demands. The Mexican people voted overwhelmingly in favor of the Zapatistas. And just as a new president took office in 2001, the Zapatistas announced their intention to send twenty-three of their commanders and the silver-tongued Subcommander Marcos to the Mexican capital. After a mobilization of millions of Mexican citizens, the Zapatistas overcame enormous political opposition by all sides of the Mexican political establishment, and the tiny indigenous woman known as Commander Ester eloquently argued the Zapatista case before the Mexican congress. Despite being the most popular piece of legislation in Mexican history, a coalition of left, center, and right establishment parties in congress flatly rejected the constitutional and legal reforms needed to implement the peace agreement previously signed by the Zapatista and the central government.

Despite the rejection of the critical Peace Accords of San Andrés by the Mexican congress, and eventually by the Supreme Court as well, the Zapatistas have continued to open schools and medical facilities throughout Chiapas. Their creative organizing and their nationalistic but enlightened social policies have vastly expanded the Zapatista bases of support inside and outside of Chiapas. I am always surprised that, despite continued repression and outright racism heaped upon the Mayans by the national government, an emotional and raw patriotism is present in every Zapatista gathering.

For example, during the anniversary celebration in late 2001, students carried a large Mexican flag to the front of the plaza and began to sing, "Mexicans, at the cry of battle prepare your swords and bridle." Interestingly, the recital of the somewhat militarist official Mexican national anthem opens every public Zapatista gathering. The national anthem continued, and the Zapatista flag remained in the background, always lower than the national banner, as adults and children sang the popular stanza, "Should a foreign enemy dare to profane your land with his soul, Think, beloved fatherland, that heaven gave you a soldier in each son." Despite military encirclements, corruption, and structural adjustment carried out by the Mexican government at the behest of the World Bank, the Zapatistas insist on remaining a proud part of the Mexican nation. They continue in the tradition of their hero and namesake, General Emiliano Zapata, who, during the Mexican revolution at the beginning of the twentieth century, placed demands upon the central government without acting to take state power by force or ballot.

On August 9, 2003, in a further demonstration of the Zapatista intent to base their governing authority in indigenous civil society, Zapatista civilian governments described as Boards of Good Government were initiated in each of five caracoles. At the same time, the Zapatista military forces announced a pullback to allow the autonomous, indigenous civilian Boards of Good Government to form their own civilian police forces and carry out all the functions of a normal civilian government.

Discovery of Transgenic Contaminations in Southern Mexico

My first discussion with Zapatistas about GMO corn took place on New Year's Day 2002. "For five hundred years our Mayan peoples have endured tremendous suffering," began the young teacher, or "education promoter" as the Zapatistas call those responsible for instructing both children and adults. "Actually there was much suffering even before the Europeans arrived, but we survived that suffering. We have survived the five hundred years since the rulers of our ancestors were defeated by Hernán Cortéz and these new rulers from Europe tried to eliminate our language, our religion, and our culture."

Seven years had passed since the Zapatistas took up arms, and it was only months after the scientific journal *Nature* published news of transgenic contamination in indigenous corn in the neighboring Mexican state of Oaxaca (Quist and Chapela 2001). Perhaps it was inevitable that discussions of the discovery of GMO contaminations in the indigenous corn of southern Mexico always included a reflection on Europe's invasion of Mexico in the early 1500s. During these early discussions, which eventually resulted in the birth of Mother Seeds in Resistance from the Lands of Chiapas, I began to sense the horror and outrage these idealistic young Tzotzil educators felt about the introduction of foreign genes into Mexican corn. In the three years since those discussions, I understand more clearly that this new invasion by transgenic corn is a total abomination, a fundamental physical and spiritual violation of the Mayan culture and of the Mayan peoples themselves.

"We must all understand that only our natural corn has allowed us to survive, to resist, to withstand our sufferings," explained another young education promoter in the summer of 2002. "That is why it is appropriate for the name of this project to be *Sme' Tzu'nubil Stzi'kel Vocol*." Another promoter picked up the train of thought and eagerly continued, "Of course, yes, *sme' tzu'nubil* means 'mother seeds' and we need that idea because our creation stories tell us we are made of corn and we renew our bodies with corn every day. For us *stzi'kel vocol* means resistance. These two words translate as withstanding suffering and that is exactly what our corn allows us to do. Our insurrection could not continue without our corn."

Can you imagine an armed uprising in which resistance is defined as "withstanding suffering"? After years of hearing Zapatista men and women use the Spanish word *resistancia* to describe their social, cultural, economic, and political program, I understood that the two Tzotzil words regularly translated as "resistancia" had nothing to do with my Western-biased image of resistance. Spanish is always a second language in the Zapatista world, and obviously this new Tzotzil translation made perfect sense: Zapatista processes have always emphasized patience, understanding, and tolerance. An oft-repeated central tenet and vision for this indigenous movement is "for a world where all the worlds fit." But I was astounded to suddenly understand that the resistance these twenty-first-century rebels envisioned utilizing to win justice, dignity, and democracy for their communities was based on *stzi'kel vocol*, withstanding suffering.

While I experienced my personal linguistic epiphany, everyone else in the small Zapatista-run restaurant in the highlands of Chiapas was focused on the Spanish-language discussion about transgenic contaminations. There were numerous side discussions in animated Tzotzil; even a popular television program was turned down to encourage participation. The apparent consensus, which I paraphrase here, revolved around the idea of *being, resisting, and persisting.* Thus, to be a Zapatista is to be in resistance; Zapatistas are prepared to withstand whatever suffering the government brings in order that there can be dignity, democracy, and justice in the world for everyone. But now they understand that governments are attacking the corn; the GMOs are causing the corn great suffering. Now it is the Zapatistas' turn to help the corn resist, because corn has always sustained their resistance.

Countless conversations such as this finally resulted in the Zapatista decision to place corn seed in long-term storage. "We know that the bad governments and the multinational corporations have stolen the people's corn seed and keep them frozen to be used to make money for the rich people. Our seed bank will be different," declared the Zapatista education official. "I don't even like the term 'bank' because we will protect our seeds without thinking that it will make us rich. We will save the corn because the corn is us and we are the corn."

Defending Corn at the Capitol

Since the time the Zapatista commanders traveled to Mexico City and spoke before Congress in 2001, the movement had entered a deep silence, and no Zapatista representative had left Chiapas or made any public statement. However, in a little-noticed but significant political act, several of the same education promoters who spoke eloquently in the New Year's discussion quoted above were sent to Mexico City to participate in a public colloquium titled "The National Forum in Defense of Mexican Corn" in 2002. As of December 2004, no other Zapatista had spoken outside of Chiapas. Because of this, I am including some pertinent portions of the Zapatista written declarations (Schools for Chiapas n.d.):

> We are people who are made of corn and earth, we are Sotsil indigenous. Today they call us Tsotsiles, because our true name was transformed on the tip of the invaders' tongue. We have been indigenous ever since our Mother Earth gave birth to us, and we shall continue to be so until that same Mother Earth engulfs us. We came to represent an Autonomous Tsotsil School, located in Oventik, Aguascalientes II, in the municipality of San Andrés Sacamch'en de los Pobres in the Los Altos region of Chiapas.
>
> It is a school that was born out of our indigenous and non-indigenous struggle, in which we sowed our struggle for an alternative education which emancipates humanity, because a people who do not know their history, their culture, are a dead people. It is a school that has no place for making distinctions about people, that is, men or women, large or small, white or dark, old man or old woman. We value everyone, and we are all valuable.
>
> We are fighting to know history, to rescue our culture. Because we are quite aware that a people who know their history shall never be condemned to repeat it, and they shall never be defeated.
>
> We have found out that the agro-chemical companies have patented our natural corn so that we will then have to buy transgenetic corn. We know the serious consequences of this type of corn they are creating, which affects our culture. For us, the indigenous, corn is sacred. If these agro-chemical companies are trying to get rid of our

corn, it is like wanting to get rid of part of our culture which we inherited from our Mayan ancestors. We know that corn is our primary and daily food, it is the base of our culture. We know that our first fathers and mothers, Tepeu and Gukumatz, created us from corn, and that is why we call ourselves men and women of corn.

That is why our grandfathers and grandmothers did not plant corn in just any way. When they planted their fields they prayed three days before Mok in our Mayan calendar, because the day Mok begins is when our indigenous grandparents surrendered their brabajos, their hearts, asking protection from our god creator so that their work would be successful. During the days of prayer they ate only three tortillas the size of a coin at three in the afternoon, and they took pozol (hominy) and dough so that the plants would not anger the earth, because our grandparents believed that the land, the trees, were beings who had souls.

We are worried that our corn will be done away with completely. That is why we want to create a seed bank in our school in order to conserve our corn, and then encourage every community to establish seed banks. A project defending our natural corn is being carried out in our school. The name of this project is "Mother Seeds in Resistance of Our Chiapaneco Land."

Safe Houses for Chiapas Corn

Early 2002, when the decision was made to place a diverse collection of corn seed in long-term freezer storage, was a time when the Zapatistas had many other pressing priorities. However, Quist and Chapela's (2001) report of transgenic contaminations in the corn of nearby Oaxaca was received with enormous alarm and urgently signaled that the very ability of the Maya people of Chiapas to continue farming was threatened.

"We have to protect these little seeds because they are under attack just like our communities," explained one young education promoter during a short break in a training session on long-term seed preservation techniques. "My grandfather was killed because he defended the traditions of our community and he believed in justice and democracy. Now, even if I am an indigenous woman, I have to defend our corn so that our traditions can continue." When we returned to the second-story classroom, large sheets of butcher paper covered the walls. In an analogy drawn from the clandestine history of the Zapatista movement, one sheet of paper boldly described in both Spanish and Tzotzil two types of "safe houses" for Mother Corn. Beside these words were two drawings. One depicted a safe house for the seed itself; the other showed many safe houses protecting the indigenous knowledge that provides both the seed and the Maya people with their eternal and interconnected cycle of life. "You see, the seed cannot survive without its people, and we cannot survive without our corn," whispered the promoter whose grandfather had been martyred.

To begin the project that would put seed into long-term storage, students who attended the Zapatista boarding school at Oventic collected highly viable, quality seed from many communities. With the help and encouragement of education promoters, the seeds that the students collected were temporarily stored in locally produced ceramic pots purchased at the nearby Sunday market in San Andrés Sacamch'en de los Pobres. Seeds were mixed with ash and lime to combat the cloud forest humidity, and a handful of eucalyptus leaves were added to each pot to ward off insects. Finally each pot was sealed with a cloth strip tied carefully around the opening.

Of course all the equipment for the seed bank could not be locally produced, and the next week saw the delivery of the freezer purchased by Martin Taylor. "What sort of a camp is this?" demanded the appliance deliveryman who had driven three hours into the mountains to deliver the freezer purchased in the state capital of Tuxtla Gutiérrez. You could literally see the mestizo driver's jaw drop as he realized that the sprawling hillside complex nestled beside a tiny Maya village was a major Zapatista center. He was incredulous when he saw that the facility included a number of large new school buildings, a large wooden auditorium covered with striking murals, an Olympic-sized basketball court facing a massive plaza, metalworking and woodworking workshops, a beautiful church, and rough wooden dormitories housing dozens of Mexican and foreign visitors in plain sight.

The delivery man's questions continued as the freezer was connected to the electrical service newly installed for this project and as a silent and dark Virgin of Guadalupe, complete with Zapatista mask, took her place on the wooden wall above the humming new white machine. "Where do all these people come from, and what are they doing here?" he asked. The indigenous community leader's only comment was to state that the hospital was taking care of patients, the school had students, and some of the people were visiting to help with projects needed by the community. I still wonder how that young man's story of his visit to Zapatista territory has evolved over the years since his return to the capital city.

"Before the seed can sleep for many years in the freezer," explained the visiting scientist, "our laboratory must verify that the moisture content of the seed is below 6 percent; otherwise, when the water inside the corn seed freezes, it will expand and burst the cell membranes, killing the seed" (see figures 7.1 and 7.2).

FIGURE 7.1. Zapatista education promoters study germination rates as a part of Mother Seeds in Resistance. Photo by Peter Brown.

FIGURE 7.2. Mother Seeds in Resistance drying tray and pot to prepare GMO-free corn for long-term storage. Photo by Peter Brown.

The education promoters set up their own production line in one of the new two-story classrooms as the day began and light streamed into the new classroom still waiting for chalkboards and electricity. One team sifted the seeds out of the lime where they were stored temporarily to keep them dry and safe from insects. Today, the red bandana masks that usually protect their individual identities while proclaiming their collective identity as Zapatistas had the more practical purpose of filtering out lime dust from their lungs.

Inside, teams of indigenous youth shuttled pots full of corn out to the sifters (see figure 7.3). Another team wrote registration numbers and collection data on the foil and plastic bags and labels and entered each collection into a central registry. The seed teams poured the corn seeds into the marked bags and took them to the drying team. There the education promoters carefully placed open bags on pans of a gypsum-drying agent inside a waterproof environment created by two large plastic bags tied with bright colored thread. Several days later, the entire group of education promoters was bashing dozens of multicolored corn seeds that balanced precariously on rocks placed on the classroom floor (see figure 7.4). "If the seed mashes when we hit it with the hammer that proves the water content is above the six percent we need," explained one teacher who happily waved a large steel hammer in one hand while balancing a

FIGURE 7.3. Mother Seeds in Resistance drying pots holding GMO-free corn seed and eucalyptus leaves. Photo by Peter Brown.

FIGURE 7.4. Zapatista corn from Chiapas, Mexico, one of the centers for the origin of corn. Photo by Peter Brown.

baby on her hip with the other. "If the seed shatters when I hit it, then the seed is dry enough to be sealed in these foil bags and placed in the freezer. There it will be safe for many years from insects as well as infection by genetically modified pollen."

Later in the day students switched to 100 percent Tzotzil as they explored the importance of corn in their communities. "I'll write it for everyone," exclaimed one enthusiastic education promoter, leaping forward. Everyone shouted out the indigenous words and spoke excitedly, all laughing and debating and talking at the same time over the finer points of using corn and the many variations among their far-flung communities. "You really are men and women of corn," joked a visiting teacher as the list of Tzotzil nouns grew longer and longer.

The hypnotic prayers of the kneeling school board members and education promoters were sung softly in Tzotzil and seemed to float lightly above the burning candles inside the school library. "We're praying for survival of the mother seeds of corn and the success of our students who have just graduated," explained the president of the school board. "With our wives and the new promoters, we ask the creator to allow this school to continue and to give us the strength to continue our resistance."

Eventually it seemed as if the prayers escaped through the metal door, gently caressed the fog-shrouded mural of school children on the front of the massive concrete library, and passed into the heavens.

Afterward, as the students walked the muddy pathways returning to their homes at the end of the semester, a tiny red spot glowed brightly outside the freezer's building signaling to anyone who cared to look that the high-tech freezer was functioning to protect the native corn seeds from GMO assault. And as moonlight streamed into the room, light from the large candles still burning in the school's library seemed to welcome the pinpoint illumination from the safely sleeping seeds. Let us all pray that these people and their corn can survive this brave new world, and let each of us who has ever eaten corn accept the heartfelt Zapatista invitation to forge new connections to the birthplace of this crop.

From the first days of Mother Seeds in Resistance, many limitations of the long-term storage program were acknowledged and discussed. Specifically, participants understood that the frozen seeds represent only a small sample of the biodiversity present in Chiapas corn and that the freezer offers a somewhat false sense of security and purity since the corn seeds that rest there have been open pollinated. Obviously it is impossible for even the mighty freezer to preserve the constantly evolving Mayan cultures, which nurture and rely on this seed. Perhaps more realistically, the "safe house" for corn gave the Zapatistas the sense that they were being dragged into the commoditization of seed and all of its administrative and technical complexities.

Chiapas Corn Goes on the Road

"We want you to take our seeds to others far away who want to give it a safe new home," declared a spokesperson for the Good Government Board in the highlands of Chiapas as all the board members nodded their heads in agreement (see figure 7.5). "This will be another way for people to know the Zapatistas, and it might help our corn to escape the dangers of transgenics, which are coming into Chiapas because of the bad governments and the big corporations."

By inviting families and farmers throughout Mexico and around the world to plant corn from Chiapas, Mother Seeds in Resistance sought to overcome some of the weaknesses associated with an exclusive focus on long-term freezer storage of seed. Solidarity growing out of GMO-free corn from Chiapas, Mexico, would allow people worldwide to become active participants in the effort to save this vital and fragile biological resource. In addition, by freely sharing their seeds with other small

FIGURE 7.5. African Biodiversity Network members receive Zapatista GMO-free corn seed for planting in Ethiopia. Photo by Peter Brown.

farmers and gardeners who agree not to claim ownership, the autonomous communities of Chiapas remind us all that life cannot be owned or patented by anyone. Ideally, a few farmers will have the technical ability to maintain genetically pure seed lines, but most solidarity growers will simply enjoy growing these beautiful corn seeds and invite friends and neighbors to a Zapatista corn feast. "No, we are not afraid that our corn or other seed is about to travel to distant lands," quietly murmured the female chairperson for the Good Government Board far in the north of Chiapas. "Many of our people have traveled a great distance, and perhaps this corn will find new homes and make us new friends. Perhaps in this way it can escape the dangers of transgenic contamination here in Chiapas. In any case, we have always shared our seed with those who need it, and we hope that others will eat and enjoy this good food" (figure 7.6).

"GMO-FREE MAYAN CORN SEED AVAILABLE FOR SANCTUARY PLANTING," boldly proclaimed the headline on the promotional flyer posted at many health food stores and local seed swaps. "Sow the seeds of resistance and join the growing movement against transgenic contamination of Mayan corn," continued the brightly colored flyer. "You

FIGURE 7.6. Via Campesina representatives receive Zapatista GMO-free corn seed for solidarity growing in Africa. Photo by Peter Brown.

can be a part of preserving a genetic heritage that has evolved over thousands of years by planting these powerful seeds in your community, farm, home, school, or family garden. Have a sanctuary corn party when you harvest and invite all of your friends and family to eat great corn while celebrating the Zapatista movement!" The flyer concluded by explaining, "Currently six types of Mayan corn are available including Highlands Purple, Highlands Yellow, Highlands White, Palenque Black, Palenque White, and Palenque Yellow. All of these seeds have been donated by Zapatista small farmers who hope that people of conscience around the world will provide respectful sanctuary for this living part of their cultural heritage."

On a windswept hill overlooking Point Reyes National Seashore several hours north of San Francisco, California, the gently swaying group of friends and neighbors faced the huge corn plants planted and nourished at the Oceansong Farm and Wilderness Center. It was just after sunset that one member of the group sang a Cherokee corn prayer from each of the four directions as everyone marveled at the beauty and strength of those plants living so far from home. "Growing this wonderful corn has been a marvelous experience," explained Benjamin Fahrer, who farms at

Oceansong without chemicals or machinery. "Next year we will plant more Zapatista corn so that it can be safe." During this first year of solidarity grow-outs, Zapatista corn has sprouted from Canada to Spain and from the Andes to Patagonia. Indigenous Pueblo farmers from New

FIGURE 7.7. Farmers at La Milpa Organica Farm in Escondido, California, USA, celebrate a harvest of Zapatista GMO-free corn.

Mexico have joined with seedsavers in Georgia, activists in Portland, *picaderos* in the Southern Cone, and schoolchildren in San Diego to provide these varieties of corn with safe new homes. Future growing seasons should see many more individuals and communities across the globe experimenting with Zapatista seeds (see figure 7.7).

Zapatista Adventures in Genetic Testing—Part 2

Genetic testing of corn samples by Zapatistas with the Mother Seeds in Resistance is not entirely new. Such testing is necessary when creating the safe houses of long-term freezer preservation of seed and when seeking safe houses in farms and gardens far from Chiapas. Each sample of corn stored in the freezer must be tested for purity, as must the seeds intended for distant grow-out by friends worldwide. The testing of saved samples of corn seed represented a defensive act designed to assure adequate reserves of genetic diversity in the face of encroaching pollution.

As the testing program for GMO contamination began in earnest, I felt a new Zapatista adventure being born. The Zapatistas were finally mounting a concrete offensive against the GMO contaminations. Individual Mayan farmers wanted to know if their seed was contaminated and were prepared to use genetic testing to plan strategies of resistance. As the Mother Seeds in Resistance from the Lands of Chiapas project matured, Zapatista representatives moved beyond the horror and outrage they first felt regarding Chapela and Quist's 2001 revelations in *Nature* about GMO contaminations in Oaxaca. This new type of testing for contamination meant that GMO contaminations were no longer unknown or invisible.

"We need to know about the resistance to these transgenics that is happening in other lands," insisted a member of the reception committee in the highlands of Chiapas during a planning meeting for the summer 2004 Zapatista corn conference. "More Zapatistas will learn the testing methods we have just observed, but please also bring people who can tell how they are responding to this new attack by the bad governments. What about Oceania and Africa? Is there resistance in Asia? Tell people we want to hear their word and will welcome them when they come here to our mountains."

Several scientists and activists did respond to this heartfelt Zapatista invitation making the summer of 2004 something of a watershed for Mother Seeds in Resistance. For the first time the global nature of the GMO threat was openly discussed in Zapatista territory. Speeches, videos, documents, and even songs transported participants around the globe quickly touching down in Oaxaca, Argentina, Africa, India, New Zealand, circling indigenous nations within the United States, and always back to the mountains of the Mexican southeast.

"Before I begin my presentation, I want to bring you greetings from the peoples of Zimbabwe," began political scientist and southern African expert Carol Thompson. "Your ancestors created this important food which is now the staple diet of the people of Zimbabwe. You will always be honored for this marvelous gift." The young and old indigenous promoters of agroecology listened with rapt attention as Dr. Thompson detailed stories of southern African resistance to genetically engineered (GE) corn. During recent drought-induced famines in Zimbabwe and other southern African nations, nongovernmental organizations discovered that the United States might attempt to introduce GE corn as food aid. As large shipments of GE corn arrived in African ports, these organizations demanded that the corn be milled before distribution. Grinding the seeds would ensure the shipments were only used as food and could not be planted and thus would not contaminate indigenous corn.

"The United States responded that grinding the corn was 'too expensive,'" explained Dr. Thompson. "Too expensive" was George Bush's reply to the African demand for grinding the GE corn. Despite the fact that the U.S. government spent millions of dollars to rain bombs upon the peoples of Afghanistan and Iraq, $30 a ton was too much to ask when it came to grinding corn shipped as food aid to southern Africa. The African governments themselves eventually responded to the popular opposition to transgenic corn seed and paid, out of their tiny national budgets, to have the corn ground before it was distributed. Later in the afternoon, Zapatista agroecology promoters were moved by the images of women in Zimbabwe harvesting and storing corn on the makeshift screen in the massive tin-roofed auditorium of Oventic.

The indigenous promoters continued their discussions in Tzotzil during detailed explanations of field testing, which were translated from Spanish by respected members of the reception committee. For this training conference, Mayan families throughout the highlands donated corn leaves from their local milpas. This new round of testing took place because Zapatistas decided that they needed to know the locations of GE contaminations. In a sense, Zapatistas needed to be able to "see" the contaminations in order to eventually attack and eliminate GE corn and GMOs in Chiapas.

Resisting the Transgenic Invasion

"Certainly we have always carefully studied the location of each military base that the bad governments sent to invade our lands." The analogy of transgenic contamination to military invasion was slowly translated into each of the languages spoken by the groups of indigenous agroecology promoters exploring procedures for the field testing of corn. "Now the transgenic contamination of our corn is a new and dangerous invasion coming into our communities. These little test kits give us a way to identify the location and strength of the bases and centers of this new invasion from the north." The dream in those misty mountains and in the steamy rainforest of the Mexican southeast is to map the distribution of GMO contamination throughout Chiapas. After identifying the location of GMO contaminations political organizing would be intensified to educate everyone in the community, regardless of political affiliation, how not to introduce additional GMO seed. "We Zapatistas have always believed that the word, our honest and true voice of our Zapatista heart, is more powerful than any weapon," commented the translator sent from the reception committee.

Eventually the communities would act together to physically challenge and eliminate GMO contamination one location at a time. Every family in a selected zone of contamination would be organized to use genetically pure seed donated from noncontaminated communities at planting time. Any volunteer plants left from seeds accidentally dropped on the ground during the previous harvest would be destroyed, and extensive testing would be undertaken before any new plants would be allowed to release their pollen into the milpa. In this manner, it is hoped that the contaminated seed will be eliminated and the field will once again produce only the genetically pure indigenous corn. Although there have been a few instances where communities in Europe have destroyed GMO test fields, no one knows if this Zapatista dream is possible or if it is scientifically feasible to reverse extensive contaminations by genetically engineered crops. However, by using a combination of Western science and indigenous insight, the Zapatistas of Chiapas have already accomplished many miracles. Perhaps their extraordinary ability to organize and the fierce loyalty of their ever-growing social base will make Chiapas the first place on earth to eliminate a full-blown GMO contamination. Perhaps people from other lands will join this battle for purity and diversity in the very birthplace of corn.

Of course, this long-range plan for making Chiapas GMO-free only makes sense in the context of global opposition to GE seed. During the summer of 2004 the Zapatistas received much evidence that such an international movement against transgenic crops does exist and may be growing. Indigenous men and women in Chiapas were repeatedly astounded to learn how far Mike, an anti-GMO activist and Greenpeace fundraiser from New Zealand, had traveled just to meet the Zapatistas. According to Mike:

> In our fight against the GMOs trying to enter New Zealand we are using a card played by our mothers and fathers when they fought to stop the gringos from putting nuclear bombs in our lands during their dirty war in Vietnam. When our national government could not find the courage to declare their opposition to nuclear proliferation, we returned to our communities. First, we asked individual families to declare their homes to be nuclear-free zones; then we went to sports facilities and churches asking them to make public declarations as nuclear-free zones.

Zapatista men and women who listened to this passionate activist were already nodding their heads in agreement. Mike continued: "When our communities spoke in unison of our opposition, our national government found its voice and outlawed nuclear weapons and nuclear power. In this way, we stopped the gringos from bringing their nuclear aircraft carriers to our ports as they traveled to bomb the people of Vietnam." Mike's closing words to Zapatista promoters, municipal leaders, and even the Good Government Board were, "Now we are doing the same thing with GMOs. We would welcome the strong voice of the Zapatistas in this struggle. Will you join with us in raising your voice against this world wide proliferation of GMO contamination?" The participating Zapatistas agreed that they wanted to declare their homes and communities to be GMO free. In the end, even the Good Government Board of the highlands of Chiapas chose to respond positively to Mike's repeated questioning of their position on the struggle again GMOs.

Toward a GMO-Free Chiapas and a GMO-Free World

The slender Tzotzil Good Government Board representative slowly tapped the side of his head with his finger, clearly smiling behind the signature red bandana. Finally he exclaimed, "Yes, yes, I have it right here. Just let me remember for a moment to be sure that I get it exactly right." After a brief pause, almost certainly searching for the Spanish words capable of summarizing the memorized Tzotzil conversation, he continued in a somewhat more serious tone:

Everyone should know that we Zapatistas are opposed to transgenics. We have participated in public forums to defend Mexican corn so our position should be clear to all. Zapatistas will always be with those who are resisting the bad governments and big corporations. In Chiapas we have instructed all of our bases of support to plant only seed which is known to them; no Zapatista will accept or plant seed from the outside.

Smiling as he gained confidence in Spanish, the Zapatista spokesperson concluded:

Everyone who is fighting transgenics should know that we Zapatistas are with them. Please tell them we are with them. Please tell them they are welcome to visit us. We in the autonomous, indigenous communities of Chiapas are resisting transgenics. Zapatistas want Chiapas to be free of transgenics and we will continue struggling until we achieve that goal.

In the silence that followed this profound and unexpected public declaration, my mind flashed to the men and women from cultures and movements around the globe who should know that the Zapatistas stand with them in resisting GMO contaminations. How could I help transmit this message around the globe? Would others recognize the significance of an indigenous statement like this directly from the birthplace of corn? Would the myriad peoples fighting GMOs look beyond the ski masks and occasional militaristic nods toward Latin American revolutionary culture to offer desperately needed support for the Zapatistas? Perhaps I can only ask that you find a way to pass along to your friends and family what you have learned and felt while reading all the way to the end of this long chapter. And I sincerely thank you for this. Perhaps someday I will meet you in the misty mountains or steamy rain forests of the Mexican southeast. I hope so. In any case. I will end in the tradition of Zapatista speakers by telling you that these are my words. I hope that you have understood them and have not been offended. These are all of my words.

Postscript

Zapatista resistance to GE crops has deepened and matured since the early investigations and actions described in this chapter, which as the book goes to press was written almost eight years ago.

Occasional field testing of corn in Zapatista communities continues today. The results from the field testing for GMO contaminations in

Zapatista communities described in the chapter are a closely held Zapatista secret. However, the expense of these testing kits, as well as the statistical difficulties of chasing the ever-changing corn genome in the high-ridged mountains and valleys of the Mexican southeast, have relegated such individual tests to secondary importance in today's resistance to GMOs.

Likewise, the early Mother Seeds in Resistance seed bank located in the highlands Zapatista center of Oventic described in this chapter has been massively eclipsed by a worldwide living seed bank of Zapatista corn. This Zapatista "seed bank" includes scientifically pure grow-outs at undisclosed locations, peasant plantings in Africa, and solidarity gardens in major cities. In fact, this living entity is now growing in dozens of countries and on every continent of the planet except Antarctica.

The movement's ecological agricultural promoters who articulate and organize the GMO resistance are themselves older and more mature. They now have years of intensive study, regular workshops, extensive training, and careful reflection under their belts. Under the guidance of these peasant educators, organic production methods have now become the norm in Zapatista communities.

GE crops have not found support even among the minority who continue to occasionally utilize chemical fertilizers, herbicides, and insecticides. In fact, the horror and anger over GMOs has extended far beyond active Zapatistas and is now deeply rooted throughout the indigenous and nonindigenous peasant communities of Chiapas, Mexico. For example, opposition to GMOs is one of the few broad social issues that indigenous Catholic, Protestant, and Evangelical activists often have in common.

Today, the indigenous Mayan peoples of Chiapas, Mexico, are increasing their policing of the source and nature of their seed stock. A Zapatista boycott of "unknown seed" includes regular community-by-community education and outreach, as well as the moral authority of the entire Zapatista movement. "We are only permitted to plant seeds we know are our friends," explained one education promoter recently. He continued on to discuss seed sharing with relatives and community members, making it clear that this sharing included respected non-Zapatistas. In fact, the Zapatista boycott of commercial seed has grown significantly throughout southern Mexico and represents a unique, massive, nongovernmental resistance to GMO contamination in today's world.

Notes

1. These are my words and my renditions of quotes from meeting with Zapatistas that were approved to be reported about. I am responsible for any errors.

References

Quist, D., and I. Chapela. 2001. Transgenic DNA introgressed into traditional maize landraces in Oaxaca, Mexico. *Nature* 414:541–543.

Schools for Chiapas. n.d. National Forum in Defense of Mexican Corn, Mexico City. January, 2002. Translated by Irelandesa. http://www.schoolsforchiapas .org/english/archive/documents/national-forum-in-defense-of-mexican-corn _.html, accessed April 24, 2013.

CHAPTER EIGHT

Preserving Soybean Diversity in Japan

RICHARD MOORE

Among industrialized countries, Japan has one of the lowest levels of food security. Although self-sufficient in rice, overall the country only produces 28 percent of its grain needs and a mere 5 percent of food and feed soybean requirements (MAFF 2005; Norin Tokei Kyokai 2003). Despite this reality, the Japanese people are strongly opposing and actively resisting the importation of genetically engineered (GE) soybeans and free-market conventional non-GE soybeans. The purpose of the resistance is to protect the biodiversity of local soybean varieties that have developed in Japan over the millennia. Soy, along with rice, is one of the most culturally significant foods in the traditional Japanese diet. This chapter examines the diversity and importance of soy foods in Japan, the nature of soy production, effects of global trade liberalization, and local resistance to the introduction of foreign varieties, especially those with genetically modified organisms (GMOs). The Japanese case demonstrates that cultural resistance combined with linking local producers to consumers can be a powerful countervailing force to the hegemony and comparative advantages of global markets.

Japanese Soybeans in Historical and Regional Contexts

Soybeans were first domesticated in China 4,000–5,000 years ago. The production of soymilk, which is used as a drink and to make tofu, started around 164 BC in the Han Dynasty of China. About 900 years later soy foods spread to Japan along with Buddhism and vegetarianism (Shurtleff and Aoyagi 1984; Liu 1997).

Traditionally, Japanese diets have centered on rice and soybeans. While rice symbolically was tied to the Shinto religion, soybeans provided protein to a diet rich in fish and vegetables. Miso soup, based on a paste made by fermenting boiled soybeans with a mixture of rice or wheat, is usually served with rice. Both miso and rice are based on regionally selected varieties adapted to differences in altitude and latitude of the narrow archipelago. As a result of the isolation of each of the 270 feudal domains in Japan during the Tokugawa Period (1603–1868), local recipes were developed that were finely tuned to local environmental and cultural differences (Shurtleff and Aoyagi 1976, 1984, 2004).

Today, the soybean continues to be a quintessential item in the Japanese diet and is rivaled only by rice in the number of ways it can be prepared. Table 8.1 shows the main soybean preparations in the Japanese diet. In 2000, the food soybean usage in Japan was 60 percent tofu, 15 percent natto, 11 percent miso, 6 percent soy protein, 5 percent soy sauce, 2 percent frozen tofu, and 1 percent soy milk. The amount in 1,000 metric tons of soybeans used for each category of soy product and the percentage supplied domestically is as follows: soybean oil, 3,616 (0 percent); tofu, 495 (15.2 percent); miso, 188 (5.9 percent); natto, 128 (10.9 percent); boiled specialty soybeans, 33 (84.8 percent); other, 174 (14.4 percent); feed, 105 (0 percent); and seed, 5 (100 percent) (Hisano 1999; Kiyomizu 2000).

Each soy food category is expressed in regional differences within Japan. For instance, natto is favored by people living in the Tohoku region (the northern six prefectures on Honshu Island). People in Sendai, the regional capital of Tohoku, also enjoy eating *zunda mochi* every *Obon*, the season when the spirits of dead ancestors come back to visit the living. *Zunda mochi* is made from crushed green soybeans mixed with glutinous rice. *Sendai* also is known for Sendai miso, a salty dark reddish miso fermented for one to three years that has subtly sweet undertones and a deep fragrance. Sendai miso has a high percentage of koji rice, the mold created by inoculating rice with *Aspergillus oryzae*, and is lower in carbohydrates. In the case of Sendai miso, the production lineage can be traced back to the individual merchants manufacturing it for the feudal lord in 1601. Shinshu miso is named after another feudal domain in what is now the area around Nagano Prefecture; it has a radiant light yellow-brown color, less rice koji, and a shorter fermentation time and is a bit more tart with a smoother texture. Consumption of soy products in Japan has gradually decreased during the last half of the 20th century. In

Table 8.1. Soybean categories in the Japanese diet

Soy Food Name	Preparation	Role in Diet
Miso	A soup paste made by fermenting boiled soybeans with rice or wheat.	Served at breakfast with rice.
Natto	Cooked soybeans are inoculated with *Bacillus natto* and incubated to make this highly digestible product with a sticky slippery texture.	Mixed with soy sauce and sometimes a raw egg, it is usually served at breakfast as a topping for rice.
Nimame	Cooked beans used in a wide array of dishes that contain sea plants such as kelp.	Used for special occasion foods, using black and green varieties.
Okara	Okara is the pulp by-product of the process of making soymilk.	Okara is used by people to make U no Hana, a dish of carrots, mushrooms, and leeks mixed with sautéed okara.
Tounyuu	Soaked soybeans are crushed and boiled. The strained liquid is served as a soymilk.	This drink is often consumed by East Asians who are lactose intolerant.
Tofu	Soymilk is brought to near boiling when the tofu (the curd) is congealed by nigari, a sea salt bittern.	Tofu can be served soft and uncooked in the summer or hard in winter.
Yuba	"Mock-goose" balls are made by deep-frying a mixture of tofu, vegetables, and nuts.	Yuba is a gourmet food associated with the ancient capital of Kyoto.
Ganmo	Ganmo is the layer that is skimmed off the top when soymilk is heated.	Ganmo is used in salads or soups typically containing sea kelp.
Kinako	The raw soybeans are ground into flour.	Soy flour is used in cookies and pastries.
Shoyuu	Shoyuu (soy sauce) is made from fermented soybeans, rice, and wheat.	Soy sauce is the most common seasoning and used on both cooked and raw foods.
Moyashi	Soy sprouts are created when soybeans are soaked in water and sprout.	Soy sprouts are used in salads and in many dishes ranging from cooked noodles to sukiyaki.

(continued)

Table 8.1. Soybean categories in the Japanese diet (continued)

Soy Food Name	Preparation	Role in Diet
Irimame	Roasted soybeans are soaked before roasting and eaten as a snack food.	On the last day of winter, families scatter them in their homes and say "Out with the ogre! In with good fortune!"
Daizu yu	Crushed soybeans.	Use as a salad oil or a cooking oil for tempura.
Edamame	Edible green soybeans— soybean pods harvested while still green.	The pods are rubbed with salt, boiled, and eaten as a summer snack.

2000, 533,000 tons of miso and 1,065,000 tons of soy sauce were produced (MAFF 2001).

Varieties, Grading, and Locality

Soy foods and local varieties in Japan were developed in different regions in response to food tastes and environmental conditions of farming. Table 8.2 shows Japan's principal soybean varieties, regions, and uses. In 2002, leading varieties were Fukuyutaka, Enrei, and Tachinagaha. As noted by Nazarea (1998), the cultural meaning of soybean production manifests in a number of production practices. Traditionally, soybean fields were upland nonirrigated fields called *hatake*. Due to small-scale farming, averaging little over one hectare with a tendency for self-sufficiency, most farms had both hatake and *ta* (paddies). Hatake were usually closer to the house, and women and elderly contributed a high proportion of the labor.

Soybeans in Japan are graded by production and consumption criteria (MAFF 2004a). The key factors are seed size, shape, weight per 100 beans, seed color, cleanliness, number of hard beans demonstrated through a germination check, and proximate composition of moisture, fat, protein, ash, and fiber, along with total free sugars, total fermentable sugars, peroxide value, acid value, free fatty acids, nitrogen solubility index, organic or genetically modified organism (GMO) content, and identity preserved (IP), which guarantees traceability. In addition, varieties are scored by

Table 8.2. Japanese varieties of soybeans (excluding edamame green edible soybeans)

Variety	Use	Production Regions
Soybeans for cooking		
Tsurumusume (Crane Daughter)	B	H
Yuuzuru (Evening Crane)	B	H
Toyomusume (Daughter of Plenty)	B, T	H
Toyokomachi (Toyo Beauty)	B, T	H
Kariyutaka (Harvest of Plenty)	B, T	H
Toyohomare (Toyo Honorable)	B, T	H
Yukihomare (Snow Honorable)	B, N, M	H
Miyagishirome (Miyagi White Eye)	B, N, O	T
Tachinagaha (Tall and Straight)	B, T	KT, TS
Ootsuru (Big Crane)	B, T, M	KT, TK, KK
Natto soybeans		
Suzuhime (Bell Princess)	N	H
Suzumaru (Bell Round)	N	H
Yukishizuka (Quiet Snow)	N	H
Suzu no Oto (Bell Sound)	N	T
Kosuzu (Small Bell)	N	T
Suzukomachi (Bell Beauty)	N	KT, TS
Natto Kotsubu (Small Natto Bean)	N	K
Suzuotome (Bell Small Girl)	N	KS
Tofu soybeans		
Oosuzu (Big Bell)	B, T	T
Okushirome (North White Eye)	T	T
Suzukari (Bell Harvest)	T	T
Tanrei (Good Looking)	T	T, HR
Ryuhou (Big and Bountiful)	B, T	T
Suzuyutaka (Bell of Plenty)	T	T, HR
Tomoyutaka (Friend of Plenty)	T	T
Tachiyutaka (Tall and Straight of Plenty)	N, T	T
Tamaurara (Round Ball Springful)	B, T	T, KT, HR
Hatayutaka (Fields of Plenty)	T	T, KT, HR
Ayakogane (Golden Woven Pattern)	T, M	T, KT, TS, HR
Houen (Glamorous)	T	T, TS, HR, KK
Nakasennari (High Yield)	T, M	KT, TS, HR
Suzukogane (Golden Bell)	T	T, HR, KK, C
Enrei (Striking Beauty)	N, T, M	TK, TS, HR, KK, C
Ginrei (Glittering Snow Peaks)	T, M	TK, TS, TK, KK, C
Nishimusume (Daughter from the West)	T	KK, C
Akishirome (Autumn White Eye)	T	TK, KK, C, S, KS
Fukuyutaka (Fortune of Plenty)	T	TK, KK, S, KS

Table 8.2. Japanese varieties of soybeans (excluding edamame green edible soybeans) (continued)

Variety	Use	Production Regions
Murayutaka (Village of Plenty)	T	KS
Sachiyutaka (Happiness of Plenty)	T	KK, C, KS
Miso soybeans		
Kitamusume (Daughter from the North)	N, M	H
Sayanami (Loaded Pods)	B, M	KT, KK, C
Tamamasari (Superior Round Ball)	B, T, M	KK, C
Tamahomare (Honorable Round Ball)	T, M	KT, TS, KK, C
Special Purpose Soybeans		
Chusei Hikarikuro (Middle Black)	B	H
Tokachikuro (Black Soybeans from Tokachi Region of Hokkaido)	B	H
Iwaikuro (Black Soybeans for Celebrating)	B	H
Tamadaikoku (Great Black Round Ball)—homonym for God of Good Fortune	B	KT, TS, HR
Yumeminori (Dreams Come True)	O	T, KT, HR
Tambakuro (black beans from Tamba)	B, S	KK, C, S
Otofuke (Otofuke Beans)	B, O, S	H
Osode no Mai (Dance with Big Sleeves)	B, O, S	H
Fukuibuki (Breath of Fortune)	T	T, KT, HR
Aomarukun (Green Rounders)	T	T
Akitamidori (Akita Green)	B, T	T
Kiyomidori (Green Beauty)	T	KS
Hayahikari (Short Season Glistening)	M, S	H
Ichihime (First Daughter)	O	T, KT
Erusuta (Large Star)	O	TK, KS

Abbreviations: Use—boiled (B), natto (N), tofu (T), miso (M), sprouts (S), and other (O); Production regions—Hokkaido (H), Tohoku (T), Kanto (KT), Tosan (TS), Hokuriku (HR), Tokai (TK), Kinki (KK), Chugoku (C), Shikoku (S), Kyushu (KS)
Source: MAFF (2004a).

taste, fragrance, and color performance in the target food categories. Because regional tastes in Japan differ, usually in a northeast to southwest gradient, local performance scores are not uniform. These factors can be responsible for drops in grade, which are costly for the producer. In 2002, the leading causes for lowering the grade of domestic soybeans were shape (35 percent), wrinkles and cracks (31 percent), discoloration (11 percent), disease (5 percent), immaturity (5 percent), and insect damage (7 percent) (MAFF 2004b). For instance, the photo in figure 8.1, taken at a taste comparison in a Japanese tofu factory visited by the author, shows a slight difference in the degree of whiteness of the tofu. The whiter block on the left was said to have more *umami* or "savoriness." Very narrow definitions used in grading can be effective in protecting local biodiversity. In the Japanese case, all soybeans have both grade and size evaluations, with larger-size beans receiving a premium. Tofu makers prefer larger beans and uniformity so that soaking time for all beans will be uniform. Domestic soybean bidding also takes into consideration where the variety was grown. For example, in 2002, large grade 1 Enrei variety grown in Toyama Prefecture fetched a 25 percent greater price than the same size and variety grown in Fukui Prefecture. However, for

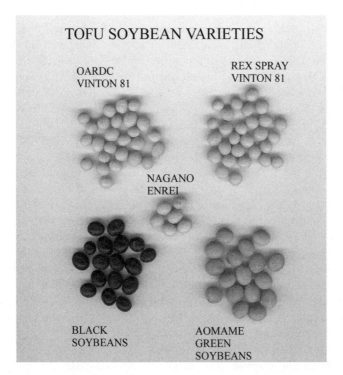

FIGURE 8.1. Lighter-colored Vinton 81 beans on the left produced better-tasting tofu with a whiter appearance than the ones on the right. Photo by Richard Moore.

the medium-size beans it was only a 5 percent premium (MAFF 2004b). The added value of the Toyama Enrei resulted from various reasons, including environmental and subvarietal differences and the effective promotion of Enrei by the Toyama Prefecture government.

Soybean Production in Japan

During the Kennedy Round of the General Agreement on Tariffs and Trade (GATT) starting in the early 1960s, foreign tariffs on soybeans decreased from 13 percent to 6 percent, and cheaper foreign soybeans, mainly from the United States, Brazil, and Canada, were imported (see figure 8.2). Prior to the drop, Japan was 28 percent self-sufficient in soybeans (MAFF 2001).[1] As noted by Yokoyama (2002), the drop in soybean production was also accompanied by decreases in domestic wheat and oat production and rapid increases in the imports for those crops.

At the same time, the Japanese government—led by the Liberal Democratic Party, which had strong farmer backing—increased subsidies for rice to the degree that many Japanese producers converted their fields into paddies. In 1970, this resulted in a rice glut. Rice (*kome*), the national train system (*kokutetsu*), and the phone system (KDD) became known as the three Ks that caused a national budget deficit. By 1970 a rice crop diversion policy was put in place that limited the amount of rice farmers were allowed to grow to 10 percent of their 1970 acreage. Farmers received subsidies for cutting their rice while it was green (*aokari*) or growing nonrice crops such as soybeans. Additional incentives to increase the scale of farming were given to groups of farmers that could combine their fields into contiguous areas. As a result of the rice crop diversion subsidy, which has grown to cover more than 25 percent of the 1970 rice acreage, many soybean growers in Japan received a 2004 subsidy of about 8,120 yen (roughly $71) per sixty-kilogram bag in addition to the price they received in the marketplace, which was approximately 4,000–6,000 yen (roughly $35–52) per sixty-kilogram bag. Calculating this into terms familiar to American farmers, a bushel (sixty pounds) of tofu soybeans in Japan earned a Japanese farmer at least $48 (subsidy included), compared with $9 that might be earned by an American farmer for the same product (non-GMO tofu soybeans).

While Japanese soybean acreage was dramatically decreasing, soybean acreage in the U.S. Midwest jumped during the 1960s. American growers and breeders selected mainly for high-yielding high-protein varieties. Indiana (I), Ohio (O), and Michigan (M), became known as the IOM region, which produced high-quality tofu soybeans. In Ohio, for

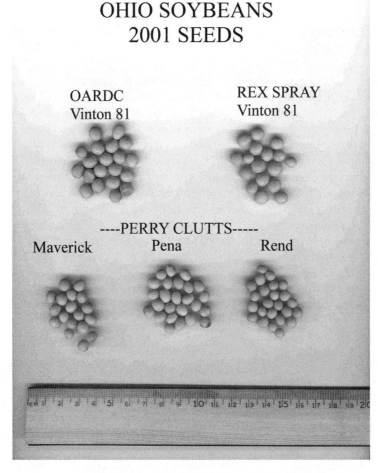

FIGURE 8.2. Soybean varieties for comparison. Photo by Richard Moore.

example, soybean acreage more than doubled from 1,419,014 acres harvested in 1959 to 3,846,614 acres by 1978 (USDA 1964, 2007). In the corn belt, soybeans used as nitrogen-fixing plants also served as a cash crop that could be rotated with corn, at a time when small dairy and livestock operations were on the decline. IOM soybeans, although excellent for making tofu, were smaller in size than the Japanese varieties.

Cultural Strategies to Protect Japanese Soybean Diversity

Cultural strategies to resist the foreign soybean varieties have occurred on several fronts in Japan. The most significant is the attempt to stop the influx of foreign soybeans by associating soybeans with the environmental aspects of rice. The government has "green-boxed" rice under World Trade Organization (WTO) rules (Fukuda et al. 2003:7). Green boxing refers to a situation where subsidies are justified if the commodity plays a key role in environmental protection or food security. In 1994, signatories of the Uruguay Round of GATT, which evolved into the WTO, created a green-box category for allowing direct payments by governments if the payments have no or very minimal effect on distorting trade. Contrasting with blue-box direct and amber-box minimally distorting price supports, the "green-box" category applies to many environmental programs that serve to strengthen the agricultural resource base. In this way, farmers can be rewarded for practicing good stewardship. Theoretically, since soybeans are usually grown as a rice diversion crop (*tensaku*), it can be coupled with the environmental benefits of rice in green boxing.

In recent years the government has also promoted the idea of the "multifunctionality" of rice, which includes environmental aspects and food security. Multifunctionality refers to the range of functions agriculture plays, including land conservation, fostering water resources, preservation of natural environment, development of favorable landscapes, maintenance of cultural heritage, recreation and relaxation, viability of rural communities, and food security.

Rice in paddies and soybeans and other vegetables in upland fields (*hatake*) were the traditional mainstay of Japan. Rice and soybeans combined the wisdom of nutritional complementarity with agroecological nutrient cycling wherein the soybeans and other cover crops such as vetch create green nitrogen fertilizer. Located in the Asian monsoon zone, Japan has an annual precipitation of 1,800 mm, almost twice as high as the world average. Japan has substantial risks of flood and water shortages mainly because of relatively steep river gradient provided by the Japanese Alps mountain range (2,000–3,000 m) that dissects the island of Honshu. Rains come in the form of seasonal torrential downpours and typhoons during the summer rainy season. (Despite the rains, Japan's high population density necessitates water conservation.) Thus, the government has emphasized to WTO that rice has environmental externalities worth more than the rice grain itself. These include preventing floods, using terraced paddies that serve as miniature dams to slow the water, channeling rainwater for irrigation and drinking water, preventing soil erosion and landslides, and promoting air purification. Comparing the value of the rice externalities ($42.8 billion) to the value of the rice itself (8.5 million tons of production times the world market price of approximately $250 per ton, which equals approximately $2 billion), we see the externalities are valued about twenty-one times the price of rice (MAFF 2004c).

The GMO Soybean Issue

Most soybeans imported to Japan come from the United States, where the rate of use of GMO soybean crops exceeds 80 percent. As a result, the GMO soybean opposition movement has taken center stage in the effort to preserve native Japanese soybean biodiversity. In the United States, GMO soybeans are synonymous with Roundup Ready (RR) soybeans produced by Monsanto, which also manufactures the Roundup herbicide. The genetics of the soybean plant were altered so that the plants are resistant to Roundup.

In 2000, GMO Starlink corn from the United States that had not been approved for human consumption was found in poultry feed and in a Japanese product for home baking sold by the Kyoritsu Shokuhin food company. In 2002, more sophisticated GMO detection techniques were developed, and trace GMO grain was found in about a third of foreign grain sold as organic. This was problematic because the organic labeling law clearly states a zero tolerance for GMO. Approximately 80 percent of Japanese consumers do not trust labels, and 70 percent of consumers strongly oppose buying GMO products. Research shows that opposition to GMO soybeans has been growing in Japan. In a study of Japanese college students, 63 percent said they were not very willing to eat food that contained GMO products, and 20 percent said they would avoid such foods (Rickertsen and Chern 2002). In fact, they were willing to pay 33–40 percent more for non-GMO soy oil, which presently is exempted from the GMO labeling law. Students in the United States scored a total of 19 percent for these categories. Research based on a representative sample of Japanese consumers from different regions, ages, and genders showed that the issue of GMO was high on the list of concerns but behind price, expiration dates, additives, nutritive content, and country of origin. When asked in 2003 about products that had GMO content, respondents listed five soy products among the top six, and 60 percent said they had concerns about eating GMO products. The greatest concern was the possible effect on the environment and potential health risks (JMAR 2004).

GMO research in Japan has been moving forward in several areas. Monsanto has been working through an associated company called Bio-Produce Group. Experiments using soy and rice have been carried out in

Ibaragi Prefecture, Hokkaido, and Akita Prefecture. One reason RR soybeans have not been as successful in Japan as in the United States is that there is more uniformity of weeds within the small soybean plots in Japan, where more hand weeding is done, particularly by smaller growers who produce the majority of soybeans. In the United States, RR soybeans had high rates of diffusion in areas characterized by large fields with different soil types (Monsanto Ohio Field Station, personal communication, 2000). One area that seems to have moderate citizen support is a new GMO rice variety that would reduce the allergy for cedar pollen, which affects many Japanese.

Since 1997 the IOM (Indiana, Ohio, and Michigan) market share for food soybean exports to Japan has declined from 1,200,000 to 1,050,000 metric tons per year (American Soybean Association 2006). At the same time, the United States' share of Vinton and other "identity-preserved" named varieties has increased from 150,000 to 400,000 metric tons per year, indicating a trend toward specialty beans, with non-IOM states participating. Identify preserved (IP) refers to the traceability of the non-GMO soybeans from the consumer through all production stages to the field of the producer. Examples of IP contracts are those that guarantee a certain "not to exceed" GMO content, that is, 99.0 percent GMO free; contracts based primarily on the International Organization for Standardization (ISO) specifying certified production and handling processes; and contracts in which production and handling activities are documented in an electronic database format (Thompson 2004). These trends demonstrate a loss of market share for IOM soybean producers as a result of the enactment of the Japanese GMO and organic labeling laws implemented in 2001.

Since 2002, groups called Soy Trust and Citizens Biotechnology Information Center have been opposing efforts by Monsanto to promote GMO research and development in Japan. The Soy Trust, started in 1998, makes arrangements with farmers to grow soybeans on their land with the production cost provided by consumer members (Hisano 1999). Harvested soybeans are distributed among consumers who provide the money. Soy Trust has successfully demonstrated and held soy parties serving locally grown soy products in areas where GMO soybean experiments have been planned in Japan. The Soy Trust is also joined by the No! GMO Campaign sponsored by the Consumers Union of Japan, the Organic Agriculture Association, and the Seikatsu (Lifestyle) Club and together they had 78,366 hectares registered as "GMO Free" areas in 2012 (No! GMO Campaign 2012). They have been successful in backing down government projects in places such as Iwate Prefecture. Japan's Ministry of Agriculture, Forestry, and Fisheries (MAFF) is currently discussing how to tighten the regulation of outdoor experimental releases of GMO crops at research centers in order to respect the Cartagena Protocol on Bio-safety, which came into force in Japan in February 2004.

Resistance through Labeling

Throughout the post-World War II period, problems have arisen with mislabeling of top varieties of rice that had been blended with the lower grades and varieties. Starting in 2000, several food labeling scandals occurred, although only a few have involved soy products. The first major scandal involved tainted milk in the Osaka area that left 14,500 people sick. Subsequently, labeling scandals have included mislabeling Australian and American beef as Japanese beef, selling imported pork from the United States as domestic Japanese pork, marketing Thai and Brazilian chicken as Japanese chicken, selling white pig pork as black pig pork, and the discovery of organic soybeans containing "negligible" amounts of GMO material. These labeling scandals, along with the mad cow disease scare, resulted in the creation of the Food Safety Commission in 2002. However, even with this new agency, consumer confidence has been shaken and remains the agency's number one concern, especially in relation to food items coming from abroad.

Under Japan's 2001 GMO labeling law, soybeans must have less than 5 percent GMO content in order to be labeled as non-GMO. As a result of this labeling law, which is more liberal than the European limit of less than 1 percent GMO content, most American non-GMO soybeans sold in Japan have less than 1–2 percent GMO content. The organic law of Japan passed in 2002 states that "organic" contains absolutely no GMO material. In 2002, a study by MAFF found that nine out of twenty-nine tofu samples using the Japan Agricultural Standards (JAS) label had "negligible" levels of GMO. The organic labeling law requires foods labeled "organic" to be GMO free. The same study also found eleven of eighteen tofu samples using the "Contains organic soybeans" label to have negligible levels of GMOs (MAFF 2003).

The importance to U.S. producers of this shift emphasizing food safety and accuracy in labeling, along with the first signs in Japan of including soybean oil under the GMO law, cannot be overstated. In 2011, the United States accounted for 62 percent of the value of soybeans imported to Japan down from 79 percent in 2007 (USSEC 2012). The United States is losing its market share of soybeans to Brazil, where national plans exist to lower the export costs of non-GMO soybeans by creating better transportation and possibly a direct link to the Pacific Ocean. As Brazil has increased its production 45 percent from 2007 to 2011 by transforming the environmentally sensitive Amazon (USSEC

2012), there may be more reason for other countries, such as the United States, to focus on the high-end environmental value-added market.

Direct marketing efforts ease the minds of consumers who have grown to have little trust in labeling. According to a survey by Japan's Prime Minister's Office in 2003, 91 percent of the respondents' top food concern was the safety of imported farm products, while 62 percent were "not concerned" about domestic food handling (MAFF 2003). Ninety-one percent of these consumers also thought that traceability was an important issue. The consumer cooperatives were the first to bridge this traceability issue with their creation of "face-to-face" relations between producers and consumers through the sanchoku (direct marketing) movement, an idea that soon spread to the competing chain stores. Trust is about 10 percent higher for the cooperatives than for the large chains. Sanchoku producer groups are usually small groups of seven to fifteen producers from the same area who market their goods together and form a relationship with consumers. The consumer cooperative fosters interactions by inviting consumers to visit with farmers on their farms and co-op stores. As a result, soy products sold at co-op stores sometimes have names, photos, or addresses of actual producers.

Much of the tofu, natto, and miso sold in Japan have labels or signs stating that it is not made from transgenic material or is made from 100 percent Japanese domestic soybeans (meaning that no GMO material was used). In addition, local labels such as the Miyagi Prefecture have found ways to promote low-input soybeans based on local conventional levels of chemical fertilizer and pesticide use (Miyagi-ken Sangyou Keizaibu [Production Economics Section of Miyagi Prefecture] 2000).

Resistance to GMOs through Buying Local

Face-to-face marketing through sanchoku was established by consumer cooperatives in the 1990s. Usually these groups are formed between the local consumers' cooperative and local groups of farmers living in the same town. The farmers' names, addresses, and even phone numbers are on the product and photos in the co-op show the group members.

In the Miyagi Co-op there was a sign showing a sanchoku group selling low-input cucumbers. The sign stated that the sanchoku team's cucumbers met the prefecture low-input certification, the skin was soft and had a fragrance like in the olden days, and that they were boxed fresh. The Miyagi Consumer Cooperative Association favors buying locally whenever possible. In the case of tofu soybeans, they prefer to give the lucrative organic soybean contracts to local producers and run the cheaper generic line made from foreign GMO soybeans side by side in the stores.

In the case of natto, the Miyagi Consumer Cooperative Association teamed with the Producers' Cooperative in Kakuda, a town in the southern part of the prefecture. The Kakuda Natto Factory uses a locally produced small heirloom bean variety named Kokosu to market beside a larger variety called Ayako.

The rise of *seikyo*, or consumer cooperatives, as the leading retailer of food in Japan is directly related to trust and concern over safe healthy local food. Today, these co-ops exist in urban and rural neighborhoods and are going head-to-head with national chain stores. The co-op law of Japan limits co-ops to activity in the forty-seven domestic prefectures. In 2005 there were a total of 626 individual co-ops situated throughout the country, operating some 1,102 retail outlets, excluding institutional and university. Co-op membership reached 23 million in 2005. Turnover continues to expand, with fresh food the main component of retail sales. Total cooperative sales in 2005 were 3,332 billion yen (US$29 billion at 115 yen/dollar), of which 47 percent was fresh food and 23 percent was dry food (Japan Consumer Cooperative Union 2006). Forty-five percent of sales were direct to consumers mainly in the form of small subneighborhood units called *han*, which serve as buying groups that put in weekly orders. *Han* is a term meaning "subgroup," and in Miyagi Consumer Cooperative's case it means at least three or four households buying together. However, the effective meaning in the co-op buying club context is that it is built upon and promotes neighborhood-level interaction often beyond that of the collective ordering. In 2005, the Miyagi co-op had the 12,516 leaders of each han meet in the early summer to discuss the situation with the consumers and the direct marketing dairy producer groups (Miyagi Seikatsu Kyodo Kumiai 2006). Individuals can have direct delivery to their homes and are also invited to attend meetings such as the above but still comprised only 16 percent of the total membership. While cooperatives in Japan are legally limited to activity within each of the forty-seven prefectures, they have created a national federation. Many co-ops were started on college campuses during the 1960s and 1970s, the founders having been active in college campus anti-Vietnam War and citizen movements at the time. The cooperatives' policy is to favor fresh, prefecturally grown, healthy, safe, and inexpensive food. As such, they have been the "watchdog" of food problems such as labeling scandals, protesting against GMO food, and pressuring the government to change policies to favor consumers.

It is also possible for producers from different nations to team together around the same commodity, related commodities, or even environmental protection to create face-to-face relations and overcome suspicions. An example is the alliance between some Ohio soybean producers and their Japanese soybean producer counterparts. In 2001 and 2002, the author took six Vinton 81 soybean producers from Ohio to Japan to discuss possible soybean export sales with the consumer cooperatives that have, as their preferential buying matrix, to buy local produce in face-to-face relations. Because there was a shortage of tofu soybeans produced locally, they were interested in creating face-to-face ties that would ensure IP traceability as well as promote ecological farming among the farmers from whom they were buying the soybeans. The co-op desired to buy organic soybeans with a higher premium from local farmers while buying the cheaper non-GMO (but not organic) soybeans from the Ohio group. At the same time, they introduced the U.S. farmers to Japanese soybean producers so that the two groups could discuss mutual interests such that the producers in the local prefecture could raise their self-sufficiency rate at the same time that both could be more aware of methods to farm more ecologically.

Lessons Learned

Japanese soybean biodiversity is still at a crossroads, and the final outcome is uncertain. However, the case does illustrate some lessons that may be applied elsewhere to preserve biodiversity.

First, it is difficult for foreign agribusiness to compete with local organizations using face-to-face farmer/consumer relations. This is an inherent problem of large-scale industrial production, especially in contexts of widespread suspicion of food products from multinational and foreign corporations.

Second, the Japanese case demonstrates great potential in marketing environmental aspects of a product, particularly in a country such as Japan where there are sharp ecotones and narrow econiches—this provides a rationale for the local population to support the commodity. The American food categories of "organic," "natural," and "conventional" do not take full advantage of the abundant opportunities for more precise labeling or the myriad possibilities for environmental and social conditions surrounding production.

Third, innovative labeling based on locality can help preserve market niches.

Fourth, citizen movements such as the Soy Trust and the consumers' cooperatives can put pressure on the local government to stop GMO research when it is not desired by the local people. Over time, it is likely that the non-GMO seed bank will become "polluted" by gene transfer or pollen from GMO varieties, even though most soybean varieties are predominantly self-pollinating. Finally, there is hope for protecting heirloom varieties through WTO green boxing. Environmental conservation payments to farmers to preserve the agricultural land base are gradually being equated with improving water and soil quality and biodiversity. It will be necessary, however, for researchers to show that heirloom varieties can be linked to other forms of biodiversity in order for green boxing subsidies to be applied to seed biodiversity.

Notes

1. In 2003, Japan's self-sufficiency rate for soybeans was approximately 5 percent overall and 15 percent for food processing varieties such as tofu soybeans. In 2001, the self-sufficiency rate for tofu soybeans was 31 percent; cooked soybeans, 88 percent; natto soybeans, 13 percent; and miso and shoyu soybeans, only 15 percent (MAFF 2004d).

References

American Soybean Association. 2006. *Soy Stats: A Reference Guide to Important Soybean Facts and Figures.* St. Louis, MO: American Soybean Association.

Fukuda, Hisao, John Dyck, and Jim Stout. 2003. Rice sector policies in Japan. USDA Electronic Outlook Report RCS0303-01. Washington, DC: U.S. Department of Agriculture, Economic Research Service.

Hisano, Shuji. 1999. Trends of Japanese soy market and new movements of farmers-consumers alignment. Proceedings of annual conference of the Agricultural Economics Society of Japan. *Journal of Rural Economics* 71:284–289.

Japan Consumer Cooperative Union. 2006. Co-op Facts and Figures, http://jccu .coop/eng/public/pdf/ff_2006.pdf, accessed May 23, 2013.

JMAR (Japan Management Efficiency Research Institute). 2004. GMO ni Kan Suru Shohisa Chosa (A Consumer Survey Regarding GMO's). Tokyo: Japan Management Efficiency Research Institute.

Kiyomizu, Tetsuro. 2000. Daizu no juukyuu doukou to kokusan daizu fukkyo no kadai [Trends in the supply and demand of soybeans and the proliferation of domestic soybeans]. *Norin Kinyuu* 10(65):671–687.

Liu, K. 1997. *Soybeans: Chemistry, Technology, and Utilization.* New York: Chapman and Hall.

MAFF (Ministry of Agriculture, Forestry, and Fisheries). 2001. *Nourinsuisan Tokei [Agricultural Statistics].* Tokyo: Agricultural Statistics Communications Department, Ministry of Agriculture, Forestry, and Fisheries.

———. 2003. *Heisei 15 Nendo Shokuryohin Shohi Monita Daiikkai Teiki Chosa Kekka no Gaiyou ni tsuite [A Summary of the Results of the First Survey Monitoring Consumer Food 2003].* Tokyo: Ministry of Agriculture, Forestry, and Fisheries.

————. 2004a. *Kokusan Daizu Hinshuu no Jiten* [*Dictionary of Domestic Soybean Varieties 2000*], http://www.maff.go.jp/soshiki/nousan/hatashin/jiten/sakuin.htm.

————. 2004b. *Kokusan Daizu no Antei Seisan no Genjo to Kadai* [*Update on the Production Stability of Domestic Soybeans*]. Tokyo: Production Division, Ministry of Agriculture, Forestry, and Fisheries.

————. 2004c. Ministry of Agriculture, Forestry, and Fisheries 2004 Environmental externalities of Japan's paddy fields farming, http://www.maff.go.jp/soshiki/kambou/Environment/env1.html, accessed September 1, 2006.

————. 2004d. Kokusan daizu no seisan ryutsu no genjo [Update on the distribution of domestic soybeans], http://www.maff.go.jp/soshiki/nousan/hatashin/daizu/kondankai/shiryou1.pdf.

————. 2005. *Shokuryou, Nogyou, Nouson Kihon Keikaku* [*Fundamental Plan on Food, Agriculture, and Farming Villages*]. Tokyo: Ministry of Agriculture, Forestry, and Fisheries.

Ministry of Finance. 2006. *Nihon Boeki Toukei* [*Japanese Trade Statistics*]. Tokyo: Ministry of Finance.

Miyagi-ken Sangyou Keizaibu [Production Economics Section of Miyagi Prefecture]. 2000. *Miyagi no Kankyou ni Yasashi Nosanbutsu Hyoshi Ninshou Seido* [*The Certification Methods for Environmentally Friendly Labeling in Miyagi Prefecture*]. Sendai: Miyagi-ken.

Miyagi Seikatsu Kyodo Kumiai (Miyagi Consumer Cooperative Association). 2006. *Sanchoku* (in English). Sendai: Miyagi Miyagi Seikatsu Kyodo Kumiai.

Nazarea, Virginia D. 1998. *Cultural Memory and Biodiversity*. Tucson: University of Arizona Press.

No! GMO Campaign. 2012. GMO Free Zone Registration Status Report (2012), http://www.gmo-free-regions.org/gmo-free-regions/japan.html.

Norin Tokei Kyokai [Agricultural Statistics Association]. 2003. Shokuryo, nogyo, noson hakusho—tsusetsu. [White paper on food, agriculture, and rural villages illustrated]. Reference Statistics Tables. Heisei 13 Nendo [2003 volume]. Tokyo: Norin Tokei Kyokai.

Rickertsen, Kyrre, and Wen Chern. 2002. Comparing consumer acceptance and willingness to pay for GMO foods in Norway and the United States. *AgBioForum* 3(4):259–267.

Shurtleff, W., and A. Aoyagi. 1976. *The Book of Miso: Savory, High-Protein Seasoning*. Berkeley, CA: Autumn Press.

————. 1984. *Tofu and Soymilk Production*. Lafayette, CA: Soyfoods Center.

————. 2004. History of soybeans and soyfoods: 1100 B.C. to the 1980s. Unpublished manuscript.

Thompson, Dennis. 2004. The Development and Evolution of IP Systems, American Soybean Association 7th Food Bean Conference, Tokyo, Japan (invited paper), http://dennisthompsonllc.com/files/Dennis%20R%20Thompson%20Publications.pdf, accessed May 24, 2013.

USDA (U.S. Census of Agriculture (Ohio)). 1964. Table 9. Acreage, Quantity, and Sales of Crops Harvested: 1930 to 1964. http://usda.mannlib.cornell.edu/usda/ AgCensusImages/1964/01/10/781/Table-09.pdf, accessed May 24, 2013.

USDA (U.S. Department of Agriculture). 2007. U.S. Census of Agriculture. Table 1. Historical Highlights: 2007 and Earlier Census Years. http://www.agcensus .usda.gov/Publications/2007/Full_Report/Volume_1,_Chapter_1_State_Level/ Ohio/st39_1_001_001.pdf, accessed May 24, 2012.

USSEC (U.S. Soybean Export Council). 2012. Japanese Trading Companies and the Asia-Bound Grain Trade. Japanese Soybean Market Intelligence: Special Report. http://www.ussec.org/wp-content/uploads/2012/10/ASA-IM -Special-Report.pdf

Yokoyama, Hidenobu. 2002. *Nihon Mugi Jukyu Seisaku Shiron* [*The History of Japanese Wheat Supply and Demand Policy*]. Tokyo: Hatsusakusha.

CHAPTER NINE

Complementarity and Conflict

In Situ *and* Ex Situ *Approaches to Conserving Plant Genetic Resources*

CARY FOWLER

Plant genetic resources, the biological foundation of the crops that feed humanity, are a living library of life. Like library books, some are conserved *ex situ* in genebanks, where they can be accessed and used for research and breeding. But not all these "books" have been collected and placed in genebanks—indeed, some have not yet been written. Professional plant breeders fashion new crop varieties, and many farmers continue to manage their crops under *in situ* conditions on the farm, selecting and altering the genetic resources under their care. The books continue to be written by such farmers and gardeners, particularly (but by no means exclusively) by those in developing countries who are harvesting, selecting, and saving their own seeds for replanting the next season.

Plant genetic resources, including seeds and other planting materials, play many different roles in society. They are an item of commerce, an indispensable element of agricultural systems, an embodiment of human creativity and culture, a product of evolution, a symbol to some of the dominance of economic and political systems, a symbol to others of resistance and defiance. They are also the subject of numerous and often conflicting claims of rights—from intellectual property rights to farmers' rights. They are the subject of international laws and debate, academic research, and dinner table discussion. They are also a living witness of how profoundly interdependent human beings and countries are.

Some 10,000–15,000 years ago, our neolithic ancestors began the process of domesticating crops (Harlan 1975). Crops have been on the move from the first day. Encounters with different cultures, and selection for different environmental conditions and human needs, created conditions conducive to the emergence of thousands of different and distinct types of rice, wheat, maize and other crops. Compared with the multimillion-year history of hunting and gathering, agriculture is but a brief moment. Agriculture's spread was thus both quick and powerful. Major crops had spread far and wide by the time the first written records appeared. The voyages of Columbus and other explorers in the fifteenth and sixteenth centuries extended and virtually completed the process by transferring crops such as wheat to the New World and maize to the Old World and beyond (Crosby 1972, 1986).

The early movements of crops within and beyond their centers of origin, and the later transfers of crops from one continent to another seem largely to have taken place "naturally" and without major controversy as a by-product of human migrations. Some exceptions are notable such as the transfer (and possible theft) of certain highly valued industrial and pharmaceutical crops such as rubber and cinchona (quinine) during the colonial era (Brockway 1979). By the beginning of the twentieth century, most crops were widely dispersed—being grown by farmers on multiple continents. By the third quarter of the twentieth century, much intraspecific genetic diversity (in the form of thousands of landraces or farmers' varieties) was also widely dispersed, situated in more than 1,300 *ex situ* collections of plant genetic resources (Plucknett et al. 1987; FAO 1998).

Looking back, we can observe that all the world's healthy and productive agricultural systems have been built on a foundation of access to both local and foreign plant genetic resources. The subject of "access," however, has become politically controversial. Charges of "biopiracy" are common. There are energetic and emotional battles over whether and under what conditions intellectual property rights can properly be claimed over plant varieties and their components. Restrictions of access are routine (though such walls may be coming down with the coming into force of yet another treaty that now promises "facilitated access"). And there are numerous and diverse attempts to extract value and benefits from plant genetic resources at every turn.

The question of access is, of course, intimately linked with the question of conservation. Access may be either from *ex situ* (genebank) or *in situ* (typically on-farm) sources. *Ex situ* systems depend on access in obvious ways. Nothing originates in a genebank. But *in situ* systems are also dependent since farmers manage genetic resources not just to conserve them but to use them for production; they continually incorporate "bred" materials from the formal seed system. This is a dynamic process. Sustaining and increasing production require access to the genetic materials that make this possible, whether in the form of modern improved varieties or additional and different farmers' varieties (landraces).

To some extent, the different approaches to conservation—genebanks and on-farm—have gotten caught up in the political controversies involving access, rights, and benefits. There are thus three "politics" of plant genetic resources: a general politics having to do with access and benefit sharing, as well as a subpolitics in which one hears a critique of *ex situ* approaches (largely from *in situ* partisans) and the opposite (largely from genebank users with scant appreciation for "traditional" agriculture).

Historically, it is safe to say that the contribution of farmers (and our neolithic ancestors) to the domestication, development and conservation of plant genetic resources, while understood (it was, after all, the subject of the first chapter of Darwin's 1859 *Origin of Species*) was not appropriately appreciated, much less rewarded. More recently, farmers' efforts have been celebrated, supported, and advanced, particularly by nongovernment organizations and academics (Brush 2000; Nazarea 1998). At the same time, some have come to view genebanks primarily (and negatively) as a servant of industrial agriculture and its commercial plant breeders. Conversely, formal sector plant breeders who access genetic resources directly from genebanks (often after extensive searches of databases containing detailed information on the individual accessions housed therein) may voice skepticism of the value of undocumented and difficult or impossible-to-access materials located on farms. Both "systems" of conservation have their drawbacks, and each serves a somewhat different immediate constituency. But, viewed from a distance, the commonalities of purpose and effect would seem to dwarf the differences.

Disparaging, or simply failing to appreciate, either the importance of farmers' contributions *or* the importance of genebanks is an act of irresponsibility, short-sightedness, and political self-righteousness in a world where threats to genetic resources found in both systems are very real. We cannot afford this—the ultimate victims will be farmers and food security. The U.N. Food and Agriculture Organization (FAO) Global Plan of Action for the Conservation and Sustainable Utilization of Plant Genetic Resources for Food and Agriculture, adopted by 150 countries in Leipzig in 1996—the first "official" recognition of the reality and importance of on-farm management and improvement of genetic resources— got it right: *in situ* and *ex situ* strategies are complementary, both are essential, and both need to be strengthened and better integrated. Both deserve and require our support. We withhold that support at our own peril.

In this chapter, I examine the ways in which global dependence regarding the conservation and utilization of genetic resources manifests— in people's diets and countries' food systems, in crop varieties, and in the flows or transfers of genetic resources. I also look at the different conservation systems referred to above and explain in more detail why both are so critical to food security efforts. Then, I describe and assess the two dominant legal conventions governing genetic resources and their effect on conservation, access, and food security. Finally, I make some observations about the current "politics" of conservation and a plea for taking a more sober and long-term view.

The Principle of Interdependence

Interdependence can be found and even measured at multiple levels from the regional and national levels down to the level of the household and even the individual crop variety. Palacios (1998) calculated the percentage of dietary food energy supplied by different foods by country and then considered whether the relevant crops were indigenous or not—in other words, whether the particular country was part of the crop's historic center of diversity. This allowed her to come up with a rough calculation of the percentage of dependence of the country on nonindigenous crops. Undertaken for FAO, Palacios's research showed that all regions— and particularly North America and Africa—are very dependent on crops that were domesticated elsewhere. Her work makes for interesting reading and even more interesting comparisons. Italy and Ghana, for example, are equally "dependent" on foods that are not native. Per capita consumption of maize (a New World crop) in Ghana was forty-one kilograms in 1992, approximately the same as the per capita consumption of tomatoes (another New World crop) in Italy (FAO 2004).

While it is true that most major crops were domesticated thousands of years ago in what is now termed the developing world, it is also true that many, perhaps most, developing countries today fall outside of all the Vavilov Centers of Origin.[1] Although it has been said that developed countries are "gene poor" while developing countries are "gene rich," the reality is far more complicated. North/South divisions blur, as perhaps they should when one is talking about events that took place thousands of years before the emergence of the nation-state. Individual countries may possess much on-farm diversity (or have a great deal in their genebanks), but this richness in a handful of crops is offset by two critical factors: (1) no country is rich in terms of the *in situ* plant genetic resources of the full range of crops important to the economy and diet of the country, and (2) while countries may have impressive diversity of certain indigenous crops, neighboring countries and regions will often contain qualitatively different and important diversity of that same crop—diversity that is unique and potentially essential for the improvement of the crop.

At the varietal level, whether in developed or developing countries, interdependence is striking. Sonalika, a once popular wheat variety (from 1966) grown around the globe that is still used as a parent in breeding programs, was itself the result of thousands of crosses involving dozens of landraces from dozens of countries. Its pedigree, or what is known of it, runs six meters long in small type, aptly demonstrating the contributions of generations of farmers, countries, continents, collectors, genebanks, and breeders to the production of a single cultivar.

Gollin (1998) examined the complete pedigrees of 1,709 modern rice varieties released by national programs (mostly in Asia) since the early 1960s. He discovered a very high level of dependency; only 145 of these 1,709 varieties were developed entirely from "own country" parents, grandparents, and other ancestors. Table 9.1 shows the extent to which released varieties rely on material from other countries, and how often a country's landraces are used by others. Evenson and Gollin (1997) have estimated that ending the cooperative exchange and breeding program in the International Network for Genetic Evaluation of Rice (INGER) would reduce the number of new rice varieties by twenty per year and entail annual economic losses of $1.9 billion.

Political positions regarding access and benefit sharing tend to be based either on historical notions about *crop* interdependency (where the crops came from in the first place, i.e., centers of origin and diversity) or *genetic resource* interdependency (the extent to which countries and cultivars are currently dependent on foreign sources of breeding materials, be those sources of Origin/Diversity or not). The world in which farmers grew only indigenous crops never really existed, at least not widely or for long. Crops moved along with the first neolithic migrations and spread. Self-reliance on one's "own" genetic resources never really existed on any scale anywhere, north, south, east, or west.

One need not dismiss, much less approve of, the early processes of globalization in the colonial era that ripped cultures apart and rendered much of the world "underdeveloped," in order to recognize and act upon contemporary needs for genetic resources in Africa, Asia, Latin America, and everywhere else—both in countries that were and were not the sites of crop domestication activities and in centers of diversity. Basing one's politics on the perceived exploitation of developing countries that occurred as crops and genetic resources spread, produces a strangely ahistorical political position—one that is based on erecting walls to the exchange of genetic resources that all countries demonstrably need. This

Table 9.1. **Summary of international flows of rice landrace ancestors, selected countries**

Country	Total landrace progenitors in all released varieties	Owned landraces	Borrowed landraces
Bangladesh	233	4	229
Brazil	460	80	380
Burma	442	31	411
China	888	157	731
India	3,917	1,559	2,358
Indonesia	463	43	420
Nepal	142	2	140
Nigeria	195	15	180
Pakistan	195	0	195
Philippines	518	34	484
Sri Lanka	386	64	322
Taiwan	20	3	17
Thailand	154	27	127
United States	325	219	106
Vietnam	517	20	497

Source: Gollin (1998).

position is pursued in the belief that prevention of "exploitation" is more important than the promotion of development. But, development without diversity is a dangerous and historically unprecedented experiment, not one to be advocated without forethought of the consequences.

One example (in addition to those cited above) will suffice, perhaps. Between 1972 and 1991, centers of the Consultative Group on International Agricultural Research (CGIAR) collected 124,000 samples from fifteen different developing countries (IFAR 1994). During the same period, they distributed 529,000 samples to those same countries, a 4-to-1 ratio. By 1992 the ratio had widened to 60-to-1 and may now be 100-to-1, in part because collecting has declined while distributions have remained high (Fowler et al. 2001). Were one to include breeding lines and other "improved" materials into the calculation, the ratio would balloon to many hundreds to one. Political/legal restrictions on exchange will never transport us back to precolonial days, but they can provide a barrier to the exit of one sample and the entry of a hundred samples into developing countries today. It would seem to be a bad trade, but unfortunately, it is one some are willing to make.

Statistics on germplasm flows into and out of countries provide an interesting but ultimately misleading (and conservative) approximation of interdependence. Regions, countries, professional plant breeders, and farmers may have an impressive amount of diversity, but they may still need exotic materials from distant lands, precisely because such materials provide traits that cannot be found in available gene pools even within highly diverse subsistence farming systems. The data provided by Gollin (1998) gives a glimpse of this. Moreover, one must consider what is actually "lost," if anything, by providing genetic resources.

Are exchanges of genetic resources really a "zero sum" game? If country A provides five samples to country B and gets three in return, is it behind by two or ahead by three? To be sure, it provided more (in numerical terms) than it received. But, now it has eight samples in total. The arithmetic of cooperation is such that virtually all countries (including developing countries) are net recipients of germplasm in today's world. An open system means that one always has access to more from the "system" than one could possibly provide to it. One's politics, therefore, is revealed by the answer to the question posed above: did country A win by three or lose by two? If the answer is that it won by three, one will want a system that facilitates access. If the answer is that it lost by two, one will want a restrictive system. But if this is the case, we will all have to live with the constraints imposed by an agricultural system with reduced evolutionary potential, reduced by the effects of decreased access to plant genetic resources.

While many of the transfers of genetic resources during the past century have gone directly into plant breeding programs, some also went straight to genebanks and to farmers for their improvement and use, especially in the United States (Fowler 1994). As of this writing, genebanks hold over 6.4 million samples (FAO 1998); 1–2 million of those samples might be considered "unique" (FAO 1998; Fowler and Hodgkin 2004). There are over 1,300 genebank collections; major ones exist in each continent (FAO 1998).

Most genebanks were established primarily to maintain working collections for plant breeders and subsequently to conserve plant genetic resources in the context of agricultural modernization and widespread shifts to modern cultivars. Many genebanks today are, unfortunately, closed-door systems: few new materials are being added to collections, and too few materials are leaving the genebank and being used by breeders. This is particularly true of many national genebanks in developing countries that were established for conservation purposes without strong links to breeding programs. Financial crises in developing countries have only exacerbated the problem, further endangering the collections.

Despite these shortcomings, the importance of genebanks to conservation efforts and, indeed, to *in situ* and on-farm management systems is quite evident. Genebanks are full of materials that no longer exist on-farm. One frequently hears about genetic erosion and the loss of genetic diversity—reason enough to promote both *in situ* conservation *and* the collection and conservation of materials in genebanks. In recent years, genebanks of the CGIAR, for instance, have restored genetic resources to at least forty countries.[2] These countries once had them, then lost them, then retrieved them from genebank collections (Systemwide Programme on Genetic Resources 1996). Had the materials not been collected and housed in genebanks in the past, they would have been unavailable and gone extinct.

In summary, as the FAO Global Plan of Action (1996) recognized, the world needs complementary *in situ* and *ex situ* conservation systems. Both systems need improvement, however, and better integration of the two is a goal to be pursued. Over 1.4 billion people live in farm families that are largely self-provisioning in terms of seeds (FAO 1998). While this *in situ* "system" deserves our respect and support, it is not a system that can quickly identify and supply needed germplasm to farmers for their own improvement efforts on anything more than a modest scale. Moreover, few *in situ* projects catalog or make genetic resources available to potential users outside of the project area. Access to genetic resources for *in situ* development efforts thus remains a problem *even* in areas and among farmers known for having impressive levels of genetic diversity. *Ex situ* efforts are similarly hampered both by political restrictions on collecting and by their poor connections with potential users in the formal and informal sectors.

A Tale of Two Laws

The Convention on Biological Diversity

Sporadic conflicts over the acquisition (and/or theft) of genetic resources have taken place literally for thousands of years. During the colonial era, some world powers attempted to control the production and spread of particularly valuable industrial crops. Food crops were rarely if ever involved, as they had spread far and wide and were typically low-cost, bulk products, hardly deserving of monopolization efforts. Botanical monopolies were difficult to establish and proved impossible to maintain even for highly prized crops such as indigo and rubber. When preserving a monopoly depends on preventing the acquisition of every single seed or cutting of a species, one can imagine that difficulties will arise.

In the twentieth century, genetic resources flowed freely in all directions. Genebanks were established, largely with materials from the south,[3] and once established, the flows reversed, with hundreds of thousands of samples heading south or circulating among developing countries. During this period, genetic resources were considered the "common heritage of mankind." The widespread distribution of genetic resources coupled with the ease of transferring and multiplying them discouraged any efforts to establish a market for genetic resources per se. Nevertheless, the importance (as opposed to the economic value, which was negligible) of plant genetic resources was becoming increasingly apparent. Indeed, the significance of biological diversity in general was receiving much attention with the rise of the environmental movement. The irony of poor countries possessing (or being the historic origin of) important botanical resources was not lost on political activists or the countries themselves. Particularly paradoxical in this context was the proliferation of intellectual property regimes that established systems of recognition (and indirectly, of compensation) for work with genetic resources that largely took place in developed countries (Fowler and Mooney 1990).

A comprehensive history of these developments remains to be written. Suffice it to say that political pressures built up to redress perceived inequities. The Convention on Biological Diversity (CBD) was one outcome of such pressures (Fowler 2001). Adopted at Rio's Earth Summit in 1992, the CBD reaffirmed countries' sovereignty over their genetic resources. Legally speaking, national sovereignty is an empty vessel—one must specify what is meant by it and must assert and enforce claims if it is to mean anything. In the case of the CBD, countries agreed that access to genetic resources must be on the basis of "prior informed consent" and "mutually agreed terms." Considering the possible options, one is tempted to say that the only alternative to this would be theft.

The CBD was based on the twin ideological assumptions that developing countries possessed something of considerable monetary value and that they could regulate access in order to establish a market and reap benefits. Significantly, many countries assumed that they had more to gain than to lose by trying to convert genetic resources into a marketable commodity. More to the point, most developing countries envisaged themselves as sellers rather than as buyers of genetic resources (a view confounded by the 100-to-1 ratio of flows of agrobiodiversity *into* developing countries cited above). At the foundation of the CBD, therefore, was a denial or at least a lack of appreciation for the reality of interdependence in the area of agrobiodiversity. In fairness, it must be noted that CBD negotiators came primarily from environmental as opposed to agricultural ministries and were perhaps influenced more by visions concerning tropical diversity of value to the pharmaceutical industry than they were by the need to maintain and facilitate flows of the rather more mundane agricultural genetic resources in support of largely agricultural economies in developing countries.

For genetic resources to be a good commodity, several criteria must be met; otherwise there is no pricing power. One must have something that is relatively if not completely unique and be the exclusive, or nearly exclusive, source of this material. Nor should there be many effective alternatives for the economically valuable qualities provided by the resource. Use or incorporation of them must provide a benefit for which customers are willing to pay a premium. One must be able to control the supply of the resource, physically and/or through intellectual property rights. In the case of traits contained in the DNA of a seed, one must be able to prevent the escape of every seed. For obvious reasons, agrobiodiversity provides few candidates that meet all these criteria. Were such conditions commonly met, nothing would have prevented the emergence of a commercial market, even before the CBD.

If, for purposes of argument, genetic resources are to become a commodity, regulated in any way by international law, one must establish what it is that falls under the assertion of national sovereignty and who has the right to regulate access or sell it. According to the CBD, genetic resources are to be provided only by countries that are "countries of origin." This means "the country which possesses those genetic resources in *in situ* conditions." This, however, does not necessarily mean the country in which a plant might be growing and be collected, because the CBD defines *in situ* conditions as "where genetic resources exist within ecosystems and natural habitats, and, in the case of domesticated or cultivated species, in the surroundings where they have developed their distinctive properties" (Convention on Biological Diversity 1993). In other words, one must first determine what the distinctive properties might be and where they first arose, and then seek prior informed consent and negotiate mutually agreed terms with *that* country, which may or may not be the country where the living material has been found. Given the likelihood that a landrace or a farmer's variety might have multiple distinctive properties, it is entirely possible that a single seed might have multiple countries of origin (Fowler 2001). Furthermore, one has the problem of distinguishing and agreeing upon what a distinctive property might be (e.g., how to distinguish between shades of red in a red apple or degrees of disease resistance) and of identifying precisely where that particular quality arose even if the event took place thousands of years ago. Following this line of argument, the problem with implementing the CBD in relation to agricultural biodiversity becomes immediately evident.

In the CBD, claims are based on geography. They are perpetual—unlike patents, they never expire. They are based on the transformation of genetic resources (what was previously deemed a common heritage) into a commodity. International laws can affect markets, but they cannot easily create them. In the case of the CBD, it appears that there were many willing sellers waiting in the wings, but few buyers. The bilateral approach simply did not work, not just because countries are so interdependent, but because identifying an "owner" and establishing a viable market is virtually impossible for biological resources like agricultural crops that have been spreading about the globe and generating diversity for thousands of years.

More than a decade after the CBD came into force, few countries can point to transactions that have netted themselves any income, but observers can point to instances in which access was denied, and biological resources lost. Correa (2003), for example, found that Andean countries provided virtually no access under the CBD regime, that most of the requests came from academic researchers rather than commercial developers like pharmaceutical and seed companies, and that virtually no benefits were generated. No access, no benefits, no winners.

International Treaty on Plant Genetic Resources for Food and Agriculture

It is important to note that the CBD does not cover collections of genetic resources assembled prior to 1992. In other words, it does not cover most existing agricultural genebank collections. CBD negotiators, instead, called on FAO to try to resolve the status of these resources, which FAO did by initiating negotiations on the International Treaty on Plant Genetic Resources for Food and Agriculture in 1994. The treaty entered into force in June 2004, following seven years of tough negotiations in which the magnetic "logic" and attraction of restrictive bilateral approaches were clearly evident even as the failure of the CBD in regards to agrobiodiversity was there for all to see.

In 1992, FAO set in motion a country-driven process to document the state of the world's plant genetic resources for food and agriculture. Based on this assessment, FAO drafted a Global Plan of Action for the Conservation and Sustainable Utilization of Plant Genetic Resources for Food and Agriculture (FAO 1996). This plan was adopted by 150 countries in 1996 in Leipzig. The original intention was for countries to adopt a new treaty simultaneously, the assumption being that the benefit-sharing component of the treaty would involve implementation of the Global Plan of Action. As noted above, however, dreams of monetary benefits permeated the atmosphere. By the time the Global Plan of Action was ready for adoption, the treaty was far from achieving consensus. It would, in fact, take another five years—during which the actions agreed upon in the Global Plan of Action faded in the consciousness of delegates.

In the final stages of negotiations for the International Treaty on Plant Genetic Resources, governments agreed that access to genetic resources would be paired, in general, with benefit sharing. Unlike the CBD, however, there would be no attempt to identify a country of origin or an owner or to link specific acts of access with specific benefits accruing to a provider. Instead, the treaty envisaged a multilateral system of access and benefit-sharing through which contracting parties would be granted "facilitated access" to plant genetic resources under the management and control of other contracting parties and in the public domain. Access to materials under development—including those under development by farmers—is at the discretion of the developer. Moreover, as the treaty recognizes existing property rights, farmers' genetic resources, not being part of the public domain, would always be at the discretion of the farmer, as they generally have been in the past.

Under the treaty (FAO 2001), monetary benefits, presumably in the form of royalties, will be paid into a financial mechanism when four conditions are met: (1) genetic resources are accessed from the multilateral system; (2) they are incorporated into a product that itself is a plant genetic resource, such as a cultivar or breeding line; (3) the product is commercialized; and (4) the product is protected by intellectual property rights in a manner that restricts further use of the product for breeding and research. In practice, these conditions will be met when a new cultivar, for instance, contains genetic material accessed from the multilateral system, and when that cultivar is protected by a utility patent system such as exists in the United States. Protection by plant breeders' rights or plant variety protection under the International Union for the Protection of New Varieties of Plants (UPOV) system[4] is not likely to trigger mandatory benefit sharing, because UPOV varieties are freely available for research and for breeding without authorization of the right's holder.

Despite the treaty, the details of the benefit-sharing arrangements remain to be worked out and approved by the Governing Body of the treaty, which held its first meeting early in 2006. Preliminary negotiations are now underway regarding the text of a Material Transfer Agreement that will accompany all plant genetic resources transferred by Contracting Parties under the treaty.

A second difficulty associated with the Material Transfer Agreement remains to be resolved—how to specify what recipients are allowed to do with the material received and how much or how little they must do to it before they are able to seek intellectual property rights over a new "invention." This issue is critical because it affects what is and is not in the public domain and thus the extent to which plant genetic resources in the treaty's multilateral system will be available to future users (Fowler et al. 2004).

A third difficulty or unknown is whether countries steeped in the culture of CBD bilaterialism will actually provide access and thus start the treaty's benefit-sharing ball rolling. The treaty mandates it, but will countries have the incentive to comply and/or the courage to enforce?

The treaty *should* reduce political tensions and thus "normalize" transfers of genetic resources—a wide cross-section of developed and developing countries have already adopted it. As, based on the foregoing argument, access is itself a benefit to all, the treaty should prove worthwhile *even* if few monetary benefits are generated by the Material Transfer Agreement. The treaty thus represents a mature and sober approach to long-standing conflicts over plant genetic resources, though it is still far from perfect.

The access and benefit-sharing provisions of the new international treaty do not cover all plant genetic resources for food and agriculture, but only a subset: approximately thirty-five crops and crop complexes and a small number of mostly temperate forages. While most major crops are included, some, such as soybean, groundnut, tomato, and most tropical forages (grasses and legumes), are excluded. Negotiations over the composition of the list of crops to be covered by the treaty's provisions on multilateral access and benefit sharing were intense, and some countries felt that they might gain more from excluding certain crops than including them. Thus, China refused to allow soybean to be placed in the multilateral system, and Latin America kept groundnut and tomato out, as well as many of the wild relatives of cassava. Latin America and Africa failed to come to an agreement over including and sharing forage legumes and forage grasses, respectively. In the future, will these countries gain enough from selling the resources to each other to offset the losses to crop improvement, development, and sustainable agriculture from discouraging integrated systems involving both grasses and legumes? Will they benefit from discouraging the breeding of disease-resistant cassava cultivars?

In the future, access to genetic resources of most crops—including materials already collected and housed in genebanks, as well as materials that might be collected or accessed in the future—will come under the legal regime of the new International Treaty on Plant Genetic Resources, at least for those countries that have ratified it. Access to genetic resources of nontreaty crops will either not be regulated by international law (e.g., in the case of materials such as soybean and groundnut assembled prior to the CBD), or will be governed by the CBD (in the case of these materials accessed after 1992). This means, in effect, that there will be considerable access barriers to genetic resources not covered by the treaty's multilateral system. With the exception of the major crops mentioned above, this will apply to so-called minor, underutilized, and orphan crops. The commercial market for seed for most of these crops is extremely small, and in most cases few if any professional breeders are working with the crop. Barriers to access to genetic resources will almost surely mean that no new breeding programs will be initiated for such crops in the future, simply because the prospect of entering into negotiations with numerous countries to assemble the genetic resources necessary to back even a modest breeding project (with limited income potential) will be prohibitive. As in the case of the forage grasses and legumes, one must wonder where the benefit lies in a system of bilateral access for such crops. Would it not make more sense to encourage rather than discourage access to and development of the genetic resources of such crops?[5]

Conclusions

As Aldo Leopold once remarked, we live in a world of wounds. Some of these wounds are to the living biological foundations of agriculture—genetic resources disappear, and are "lost" forever. Genetic resources are so central to agriculture, to food security, to development, to business and commerce, to human cultures and our own history, that it is little wonder that they have become both the subject of much interest and fascination and the object of considerable political controversy. One thing seems clear: struggles over plant genetic resources will not resolve the underlying problems that motivate and energize many of the actors involved in the disputes, for no matter how important these resources are, they simply do not have the power to redress the many problems and grievances to which they may be linked.

In the end, we must get back to the beginning. We must realize that these resources are, in fact, a precious gift bequeathed to us by our ancestors. Whether we now live in a developed or developing country, in a city or on a farm, go back a few generations and we find that we are all farmers.

Much of the politics of genetic resources of the past twenty-five years has revolved around two central issues: (1) stopping the appropriation of a public good by private interests (chiefly through intellectual property rights) and (2) encouraging governments to take responsibility for the conservation of resources, that are, by their history, a "common heritage" for all of us. Neither of these political battlefronts has yielded many victories.

Victories are not the same thing as solutions, as Nobel laureate David Hume once remarked. The time has come to look for solutions, starting with ways to ensure the conservation of the very resources that are disappearing as we argue over how to save them. We no longer live in neolithic times or in Vavilov's world demarcated by nonpermeable Centers of Origin. We live in a highly interdependent world, one in which the major crop (maize) of the poorest region (Africa) is a botanical immigrant from the Americas. It would seem self-evident that while Africa may be a supplier of some resources, it has to be a recipient of others. Long-term cooperation and sharing are indicated, particularly in light of the overwhelming obstacles to creating, enforcing, and benefiting from a closed system in which the resources can be bought and sold, or hoarded and lost.

The evolution and future of food crops is in our hands. It behooves us, therefore, to construct a political framework that facilitates the conservation of the raw materials of this evolution. Genetic resources exist both *in situ* (principally on farm) and *ex situ* (principally in genebanks). Genebanks, of course, are stocked with materials accessed from farmers. But, and this must be recognized, genebanks are also, today, stocked with materials that have been lost from *in situ* systems. Both systems— *in situ* and *ex situ*—contain materials the other does not have. All of these resources should be conserved regardless of the conservation regime they are currently under. As we know, neither system is perfect. Losses occur in both. The rise of on-farm projects celebrating and promoting management and development at the farm or homegarden level is encouraging. Equally encouraging are efforts to improve and rationalize genebank operations and to secure funding (through the Global Crop Diversity Trust) to ensure permanent maintenance of collections. Proponents of each system should be active and vocal supporters of the other. If they are not, they have missed the point, and something is dreadfully wrong. In sum, support of one conservation method does not require, nor is it rendered more effective by, opposition to the other form.

While historic flows of germplasm have played an important role in the development of agriculture everywhere, *current* flows are indicative of *current* needs. As noted above, current transfers of genetic resources are numerically large and in virtually all cases each country is a net recipient. More important, each country has access to materials from others that are unavailable locally. While it is true that different countries have different capacities to use and exploit these materials, it is also true that (1) regulation or restriction of access will not substantially alter this fact, and (2) restrictions will disproportionately harm those with reduced capacity, that is, developing countries that desperately need genetic (as opposed to chemical or other) solutions to otherwise intractable problems (e.g., disease resistance). Even farmers who save their own seeds and manage impressive diversity in their own fields are, like professional breeders, arguably in need of genetic resources in order to introduce new traits and improve their crops, particularly in the face of climate change. Thus, both farmers and breeders depend on a conducive political and legal environment that facilitates flow of genetic resources within and between *in situ* and *ex situ* systems and between countries. While international law can help create part of this environment, much depends on decisions made at the working level by countries, by the private sector, by organizations (including civil society groups), and by individuals. Practicing respect, building trust, and behaving in good faith with current and future generations are vital ingredients to fostering unhampered exchange, research, and development.

Notes

1. The Russian scientist, N.I. Vavilov, working in the first half of the twentieth century, studied the geographic distribution of crop diversity and mapped out what he described as the crops' "Centers of Origin," or more properly centers of genetic diversity (Vavilov 1949).

2. These countries are Afghanistan, Argentina, Bolivia, Botswana, Brazil, Cambodia, Cameroon, Chile, Dominican Republic, Ecuador, Eritrea, Ethiopia, Gambia, Guatemala, Guinea, Guinea-Bissau, Honduras, India, Iran, Iraq, Kenya, Liberia, Mali, Mexico, Myanmar, Nepal, Nigeria, Pakistan, Panama, Paraguay, Peru, Philippines, Rwanda, Senegal, Sri Lanka, Sudan, Tanzania, Turkey, Uruguay, and Zambia.

3. Though collections in developing countries house an impressive number of "genuine" landraces of local origin, farmers' varieties developed over a long period of time, for instance, maize varieties developed by U.S. farmers from the seventeenth to the mid-twentieth century.

4. UPOV is an organization that oversees and works to harmonize plant variety protection laws worldwide. It consists of countries that offer such intellectual property rights for new plant varieties.

5. For a more comprehensive analysis of the treaty, see Fowler 2004.

References

Brockway, L. 1979. *Science and Colonial Expansion: The Role of the British Royal Botanic Gardens*. New York: Academic Press.

Brush, Stephen B., ed. 2000. *Genes in the Field: On-Farm Conservation of Crop Diversity*. Rome, ON: International Development Research Centre (Canada); International Plant Genetic Resources Institute.

Convention on Biological Diversity. 1993. Article 2, Use of Terms, http://www.cbd .int/convention/articles/default.shtml?a=cbd-02, accessed May 29, 2013.

Correa, C. 2003. The access regime and the implementation of the FAO international treaty on plant genetic resources in the Andean group countries. *Journal of World Intellectual Property Rights* 6:795–806.

Crosby, A. 1972. *The Columbian Exchange: Biological and Cultural Consequences of 1492*. Westport, CT: Greenwood Press.

———. 1986. *Ecological Imperialism: The Biological Expansion of Europe, 900–1900*. Cambridge: Cambridge University Press.

Darwin, Charles. 1859. *On the Origin of Species by Means of Natural Selection, or the Preservation of Favoured Races in the Struggle for Life*. London: John Murray.

Evenson, R. E., and D. Gollin. 1997. Genetic resources, international organizations, and improvement of rice varieties. *Economic Development and Cultural Change* 45(3):471–500.

FAO. 1996. *Global Plan of Action for the Conservation and Sustainable Utilization of Plant Genetic Resources for Food and Agriculture*. Rome: U.N. Food and Agriculture Organization.

———. 1998. *The State of the World's Plant Genetic Resources for Food and Agriculture*. Rome: U.N. Food and Agriculture Organization.

———. 2001. *International Treaty on Plant Genetic Resources for Food and Agriculture*. Rome: U.N. Food and Agriculture Organization.

———. 2004. FAOSTAT database, http://faostat.fao.org, accessed October 22, 2005.

Fowler, C. 1994. *Unnatural Selection: Technology, Politics and Plant Evolution*. Yverdon, Switzerland: Gordon and Breach.

———. 2001. Protecting farmer innovation: The Convention on Biological Diversity and the question of origin. *Jurimetrics* 41(4):477–488.

———. 2004. Accessing genetic resources: International law establishes multilateral system. *Genetic Resources and Crop Evolution* 51(6):609–620.

Fowler, C., G. Hawtin, R. Ortiz, M. Iwanaga, and J. Engels. 2004. The question of derivatives: Promoting use and ensuring availability of non-proprietary plant genetic resources. *Journal of World Intellectual Property* 7(5):641–664.

Fowler, C., and T. Hodgkin. 2004. Plant genetic resources for food and agriculture: Assessing global availability. *Annual Review of Environment and Resources* 29:143–179.

Fowler, C., and P. Mooney. 1990. *Shattering: Food, Politics, and the Loss of Genetic Diversity*. Tucson: University of Arizona Press.

Fowler, C., M. Smale, and S. Gaiji. 2001. Unequal exchange?: Recent transfers of agricultural resources and their implications for developing countries. *Development Policy Review* 19(2):181–204.

Gollin, D. 1998. Valuing farmers' rights. In *Agricultural Values of Plant Genetic Resources*, edited by R. E. Evenson, D. Gollin, and V. Santaniello, 233–245. Wallingford, UK: CABI.

Harlan, J. 1975. *Crops and Man*. Madison, WI: American Society of Agronomy/ Crop Science Society of America.

IFAR. 1994. *Agriculture in (Name of Country): The Role of International Agricultural Research Centres*. Arlington, VA: International Fund for Agricultural Research.

Nazarea, V. D. 1998. *Cultural Memory and Biodiversity*. Tucson: University of Arizona Press.

Palacios, X. F. 1998. Contribution to the estimation of countries' interdependence in the area of plant genetic resources. Background Study Paper 7, Rev. 1. Rome: U.N. Food and Agriculture Organization, Commission on Genetic Resources for Food and Agriculture.

Plucknett, D. L., N. J. H. Smith, J. T. Williams, and N. M. Anishetty. 1987. *Genebanks and the World's Food*. Princeton, NJ: Princeton University Press.

Systemwide Programme on Genetic Resources, Consultative Group on International Agricultural Research. 1996. *Report of the Internally Commissioned External Review of the CGIAR Genebank Operations*. Rome: International Plant Genetic Resources Institute.

Vavilov, N. I. 1949. *The Origin, Variation, Immunity and Breeding of Cultivated Plants*. Waltham, MA: Chronica Botanica.

CHAPTER TEN

Situated Meanings of Key Concepts in the Regulation of Plant Genetic Resources

KRISTINE SKARBØ

The stakes in the realm of agricultural biodiversity are high: rural live-lihoods around the planet, income and incentives in agricultural indus-tries, and the very foundation for future food supply. Biodiversity connects actors in all corners of the world with common and competing interests into networks of transfer and exchange (Escobar 1998). The last decades have witnessed budding dreams about—and numbing con-flicts over—the potentially profitable derivatives of plant genetic resources (PGR). Diverse actors and interests have given rise to tense debates about the rights to these resources and the distribution of benefits from their use. Amidst this heated discussion a number of initiatives aiming to regulate the realm have emerged.

Early regimes to regulate PGR were conceived in the era when plant breeding was becoming an industry and focused on plant variety protec-tion as an incentive for breeders. Other concerns have since been added to the agenda of these regulations, including the disappearing diversity of landraces from farmers' fields (Frankel 1973; Fowler and Mooney 1990) and the lack of compensation along with the threat of losing access to germplasm for farmers, especially those in the Global South (Shiva 1993, 1997). New regulatory frameworks have been proposed during the last two decades recognizing that these custodians and original developers of landraces should be compensated for their use. In 1989, farmers' rights were introduced as a progressive counterpart to plant breeders' rights in the International Undertaking on Plant Genetic Resources for Food and Agriculture of the U.N. Food and Agriculture Organization (FAO 1983; FAO 1989). The Convention on Biological Diversity (CBD; effective 1993) declared that nations have sovereign rights over their biodiversity and that careful negotiations, encompassing prior informed consent, mutually agreed terms, and access and benefit sharing, must occur before any plant material transfer (UNEP 1993). Partly as a response to the inconvenience and high transaction cost these complicated negotiations would have for the particular instances of crop genetic resource use, the International Treaty on Plant Genetic Resources for Food and Agriculture (effective 2004) established a multilateral system for facilitated exchange and access to planting material for a number of important food and fodder crops (FAO 2001; see also Fowler et al. 2001 and chapter 9, this volume). The treaty also elaborates on access and benefit sharing and the protection of farmers' rights.

Thus far, however, implementing these ground-breaking treaties has proven elusive (see chapter 9, this volume), and examples of successful biodiversity transfers with benefit sharing under the new frameworks' conditions remain few (Brush and Stabinsky 1996; Petit et al. 2001; Hayden 2003; Srinivasan 2003; Rosenthal 2006; Wynberg et al. 2009). The most promising development is increased access to the genetic resources of sixty-four crops under the International Treaty on Plant Genetic Resources. Five years after this policy was effected, the first round of grant distribution from its benefit sharing fund began. In 2009, eleven initiatives to strengthen the maintenance of biodiversity in developing countries were supported by these funds, including the site at which I conducted the research presented here (FAO 2009). Yet even if these important steps are being taken in the right direction, unresolved issues regarding the regulation of rights to and benefits from biodiversity continue to restrict exchange of much of these crucial resources. Since global negotiations inherently involve stakeholders with myriad perspectives, to have agreed on these treaty texts is a major advancement. But full implementation is wanting, and divergence in the interpretations of what the key concepts from these texts signify has been identified as one problem that may complicate this process (Posey 1994; Cleveland and Murray 1997).

This chapter explores how key concepts in the debate about conserving and regulating PGR are interpreted by different actors. I use a relatively new research technique called "photovoice"—a promising photographic elicitation method that hands over the camera to participants—to document and understand the differences in how biodiversity, rights, and benefits are perceived by groups of people with different stakes and roles in relation to PGR. As we shall see, this method is a powerful—and empowering—tool to gain understanding of the multiple perspectives that need to be taken into account when working out issues of implementation of these initiatives so important to our planet's diversity of life.

Situated Meanings

The importance of people's points of view in shaping the ways we process, represent, and act upon the things we perceive has been a focus of anthropology and other disciplines in recent decades. Ethnoecologists have shown that people have extensive knowledge of their local environments. Classification studies (Berlin 1992) and research linking local or indigenous knowledge to interaction with landscapes and natural resources (Nazarea 1999a; Berkes 1999) have shed light on human cognition of their environment and the relationship between cognition and behavior. Although contested, indigenous knowledge has gained prominence as a new paradigm in development, representing a crucial resource for living in and adapting to new situations in localized places around the world (Agrawal 1995; Orlove and Brush 1996; Sillitoe 1998; Ellen 2007). However, parties other than indigenous people have their own representations that matter, too. Different frames of interpretation can pose problems if they are ignored, as when modern agricultural technology is imposed in a place without considering existing systems (Lansing 2006), or when researchers firmly, but perhaps wrongly, believe their agenda and results are unattached to their own vantage point (Bourdieu 1987; Marcus and Fischer 1986; Haraway 1988). However, if it is recognized, the situated knowledge of farmers, scientists, or any other group of people can be illuminating—a promising pathway to deeper understandings and new solutions (Nazarea 1999b). This insight guides the present search into different actors' situated understandings of the key concepts that frequently appear in legal conventions concerning PGR. If these differences are recognized and acknowledged, they can become bridges among people and communities with a broad range of backgrounds.

Three Players in Conservation

There are three primary groups of people that are working with the management of potato biodiversity in Peru: farmers, nongovernmental organization (NGO) employees, and scientists. These three significant and distinct players in the biodiversity game represent bodies that collectively manage some of the world's largest *ex situ* and *in situ* repositories of potato genetic resources, right in the heart of the crop's center of origin and diversity. They are tied together through a novel conservation initiative in Pisaq, a county in the department of Cusco in the southern Peruvian Andes (see figure 10.1). Here six communities have joined to establish a biocultural conservation area, the Parque de la Papa (Potato Park), assisted by a regional NGO, the Asociación para la Naturaleza y el Desarrollo Sostenible (ANDES; Association for Nature and Sustainable Development). In 2004, representatives from the Parque de la Papa signed an agreement with the Centro Internacional de la Papa (CIP; International Potato Center), an agricultural research center and *ex situ* genebank, for the repatriation of potato landraces collected from surrounding areas (ANDES, Parque de la Papa, and CIP 2004). To date, over four hundred varieties have been transferred from CIP's genebank to local Quechua farmers.

The six communities in Pisaq that constitute the Parque de la Papa are located in the Vilcanota Valley east of the departmental capital of Cusco and cover an area of about 12,000 hectares (CIP n.d.-a.). ANDES-supported activities in the park are centered on the protection of native

FIGURE IO.I. Map of study sites in Peru.

agrobiodiversity. The formal repatriation of native potato varieties from CIP along with the signed treaty are major institutional accomplishments receiving international attention (*New Scientist* 2005; Suri 2005; Salazar 2008). The Quechua farmers in this area carry out agriculture with relatively limited mechanization. In addition to potatoes, they grow native varieties of other roots and tubers, along with corn, quinoa (*Chenopodium quinoa*), kiwicha (*Amaranthus caudatus*), cucurbits, grains such as wheat and barley, and vegetables such as cabbage and carrots. Since the communities in the park are located at different altitudes, the assortment of crops grown in each varies, with potatoes planted farthest toward the mountain peaks. The crucial role of the farming families in sustaining diversity speaks for itself—any framework of regulation intended to enhance on-farm cultivation of native diversity will not work unless it works for them.

ANDES, founded in 1995, describes itself as "a non-profit Peruvian indigenous organization which aims to improve the quality of life of Andean indigenous communities by promoting the conservation and sustainable use of their bio-cultural heritage through rights-based conservation-development approaches" (ANDES n.d.). The organization has its main office in Cusco but works with projects in various parts of the southern Peruvian Andes. Most of the employees are recruited locally, although some staff and volunteers come from other parts of the country and abroad. ANDES thus has solid roots in Peru but is networking with and receiving donor support from institutions in northern countries. NGOs today play major roles in shaping national and global civil societies (Tvedt 2002). ANDES is one of a growing number of southern NGOs, whose engagement is expected to become increasingly important in the future (Van Rooy 2000). In Latin America, NGOs have grown in prominence as social actors during the last decades as states withdraw from the public arena and leave tasks such as management of national parks and agricultural extension to private institutions (Bebbington 1997). In the context of these global and regional developments, we can appreciate the potentially important role of organizations such as ANDES in the implementation of PGR regulation.

CIP, founded in 1971, "seeks to reduce poverty and achieve food security on a sustained basis in developing countries through scientific research and related activities on potato, sweet potato, and other root and tuber crops and on the improved management of natural resources in the Andes and other mountain areas" (CIP n.d.-b.). CIP is based in Peru's capital, Lima, and is part of the global Consultative Group on International Agricultural Research (CGIAR) system. It is thus one in a network of fifteen international research centers spread around the globe. These centers have played a major role in agricultural development during the last four decades through collection, storage, breeding, and distribution of crop varieties, along with other mandated research and extension activities (Hawkes 1985; Plucknett 1987; Lele and CGIAR 2004). Established as the research and development arm of the Green Revolution, the mandates and agenda of the CGIAR and its associated centers have shifted over time (CGIAR n.d.-a). Now sustainable food production is the main focus, and, contrary to the case in the 1970s, reduced pesticide use is promoted through integrated pest management. In what has been referred to as "the greening of the centers," the present mission aims to "achieve sustainable food security and reduce poverty in developing countries through scientific research and research-related activities in the fields of agriculture, forestry, fisheries, policy, and environment" (CGIAR n.d.-b). Each center primarily works with a certain set of crops or farming systems. CIP not only coordinates research and extension in root and tuber crops but also manages a genebank collection boasting over 3,800 accessions of cultivated potatoes native to the Andean region. Several hundred of these varieties are now being repatriated to the Parque de la Papa (see chapter 1, this volume). With one of the world's largest tuber collections and institutional ties across the world, CIP continues to play a major role in the implementation of PGR regulations.

Research Design and Methodology

In order to gain insight into how farmers, advocates, and scientists interpret key concepts from PGR regulation, a relatively new method called photovoice was employed. Originally developed to capture and articulate views of people commonly underrepresented in official communication arenas, photovoice involves participant photography combined with participants' own interpretations and explanations of their images (Wang and Burris 1997; McIntyre 2003). A photograph can be a powerful communicator, literally providing a picture from the photographer's point of view. However, subjective interpretations of the same photograph often vary, and thus the photographs need to be accompanied by an explanation in order for their intended message to come through. This exercise provides an opportunity for the participating photographers to reflect upon particular topics, and during discussions new insights may arise (Warren 2005).

Whereas most prior photovoice projects have focused on bringing to the scene the voices of one particular, often marginalized, group, this study adds a new dimension to the methodology by using the technique to bring forth the voices of different stakeholders on the same problem. Participants representing groups that typically communicate through exclusive channels are in this case standing on equal ground and use the same form of expression—photographs and oral explanations thereof. The three concepts forming the focus of the study and of the photographs are biodiversity (*biodiversidad*), rights (*derechos*), and benefits (*beneficios*). These were chosen because of their central place in regulatory texts and the debates that surround the regulation of PGR.

Data Collection and Analysis

Data collection took place between May and August of 2006, in Pisaq, Cusco, and Lima. During May and June, participatory observation in the three locations allowed for the identification of potential participants, to whom the project was explained and invitations to participate extended. A total of twenty-five disposable cameras were then distributed among farmers in the Parque de la Papa (n = 10), ANDES employees (n = 10), and CIP scientists (n = 5), with instructions to take pictures illustrating biodiversity, rights, and benefits. Each camera had twenty-seven exposures, and within this upper limit, the participants were encouraged to take as many pictures as they wanted to illustrate each concept. In the end of July, cameras were collected, film developed, and all pictures brought back to their respective photographers. Out of the twenty-five distributed cameras, fifteen were returned with pictures. Two of the participating scientists chose not to use the cameras but instead provided digital photographs that either they or their colleagues had taken. Focus group discussions and individual interviews were arranged for the photographers to explain their pictures and the meanings of the concepts. A total of eight farmers, six NGO workers, and three scientists contributed photographs to the study, and everyone who contributed photos participated in discussions, except three of the farmers, whose photographs instead were commented on by other participating farmers from their communities. All interviews were conducted by the researcher in Spanish, except for one scientist interview and discussion that took place in a combination of English and Spanish.

As exciting, engaging, and empowering for both researchers and participants as this new photographic method may be, its recent creation means that ways of analyzing the data are still not thoroughly developed. In fact, most of the academic literature, where it appears, focuses more on the method per se than discussion of research results. In this study, I have experimented with different styles of analysis and interpretation that emerged during the research process in the hope that this work can serve as stimulus for further development of the technique.

All along, discussions and interviews were tape-recorded, and these transcripts formed the basis for analysis and interpretation together with the photographs. ATLAS.ti 5.0 was used to create a database with all transcripts and photographs. The sections of the transcripts describing or discussing the photographs were identified as "quotations" and coded manually. Quotations, codes, and photographs were combined into "networks" that formed units of analysis. Multiple shots of the same scene by the same photographer were compiled into the same network, so that from the total of 163 photographs that were discussed, a final 117 networks containing either single or several photos were compiled (see figure 10.2).

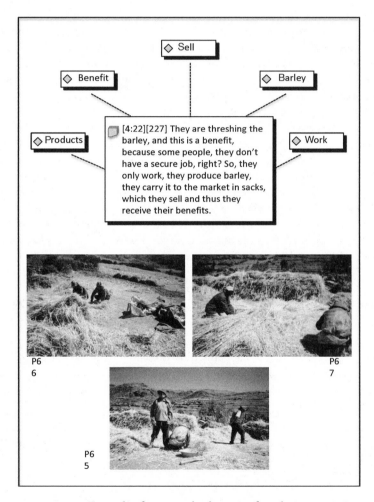

FIGURE 10.2. Example of a network, the unit of analysis, consisting of photos of a situation, a quotation with the photographer's explanation, and codes extracted by the researcher.

Concept Distribution and Scene Choice

As shown in table 10.1, the concepts of rights, benefits, and biodiversity were prioritized somewhat unevenly across the different groups of research participants. All participants were similarly instructed to capture photo illustrations of each of the three concepts, but no specific guideline was given for how many to take of each, except for the total limit set by the camera (twenty-seven exposures). Chi-square tests indicate that both on a general level and within the three groups, people have paid more attention to some of the concepts than others ($p < 0.001$ in all cases). Farmers paid attention to benefits and rights on a nearly equal level but took fewer photographs illustrating biodiversity, despite the fact that biodiversity is extremely relevant to their livelihoods and salient in their landscape. Moreover, it might arguably be something more concrete and thus easier to capture in a photograph. This indicates perhaps not that biodiversity is unimportant for them but that it is *just there*, and from their perspective, benefits and rights are topics that are more pertinent to discuss than the former. NGO workers focused more attention on rights, less on biodiversity, and least on benefits. This may reflect their daily agenda since ANDES explicitly focuses on promoting farmers' rights and biodiversity conservation. Scientists clearly devoted the most attention to benefits, which comprise 70 percent of their illustrations. This may be interpreted as a reflection of the contemporary international concern that benefits be brought back to the farmers who are the primary sources of germplasm. Scientists' photos tend to emphasize the benefits farmers derive from the research and development work of CIP. From this overview, we may observe that different terms carry different degrees of salience for different actors, reflective of their situation.

Further, differences appeared regarding the scenes and situations in which participants chose to locate their photographs and discussions.

Table 10.1. Frequency of concepts illustrated

Participant group	N	Number of photo networks	Concept		
			Rights	Benefits	Biodiversity
Farmers	8/5[a]	59	42%	47%	10%
NGO employees	6	31	52%	19%	29%
Scientists	3	27	15%	70%	15%
Total	17/14	117	38%	45%	16%

[a]Eight farmers contributed photos; only five participated in interviews/discussions.

Farmers and NGO workers used the cameras both during their professional and private lives, whereas scientists, with one single exception, took pictures exclusively on professional occasions. Discussions followed a similar pattern. Farmers and NGO workers presented a wide range of concept illustrations and talked about how biodiversity, rights, and benefits played roles in their own personal lives. For example, they gave examples involving themselves or their families and neighbors, their homes, and personal hikes, travels, and events. Scientists tied most of their photographs to conventions and regulations of the PGR sector and did not talk about the significance of biodiversity, rights, and benefits to them personally. Their images were mostly taken on visits to rural communities in relation to their work with CIP and in no case involved situations from their private lives. These differences remind us of the different roles the stakeholders play in relation to biodiversity. For farmers there is no sharp distinction between a private and a professional sphere—biodiversity management and its attached rights and benefits concern them not only professionally as farmers, but also as individuals, families, and communities. For scientists, on the other hand, these issues are to a greater degree linked with professional activities, separate from the non-work-related aspects of their lives.

Beginning with biodiversity, in the following I discuss the ways in which each of the concepts was illustrated by participants through their photographs and explanations.

Representing Biodiversity

Biodiversity. Everything, everyone that has life is this. For example the plants, right? The animals have life. People also have life, right? Everything we have got that lives, has life, right?

Farmer, Parque de la Papa

Biodiversity, for me, theoretically, is everything, right? It is life, it is existence, it is creation, it is this, it is energy.

NGO employee, ANDES

Not only plants, animals, microorganisms, but the human being as well, right? I think that we cannot isolate ourselves from this whole sector of the ecosystem, from animals, from living beings.

Scientist, CIP

There are striking similarities among the above explanations of biodiversity. They all agree that biodiversity refers to living beings, "everything that has life." Their statements also conform with the most central definition of biological diversity in the context of international treaties, from

a

b

FIGURE 10.3. Picturing biodiversity. (a) Scientist: "The biodiversity of potatoes allows harvests, because they have different levels of resistance or tolerance, and when cultivated in mixture they interact to increase the average resistance level" (René Gomez, CIP). (b) NGO employee: "Here is a garden. These are medicinal plants. This is the nettle and that is the yacón [*Smallanthus sonchifolius*]. You see that they grow here too, right? That one can grow them in the house as well" (Yanina Usucachi, ANDES). (c) Farmer: "Biodiversity is for example, the plants. Here you see everything, right? But people are missing" (Justino Yuccra, Parque de la Papa).

c

FIGURE 10.3. (continued)

the CBD: "The variability among living organisms from all sources including, *inter alia*, terrestrial, marine and other aquatic ecosystems and the ecological complexes of which they are part: this includes diversity within species, between species and of ecosystems" (UNEP 1993:3).

The similarities among different stakeholders in their basic conceptions of biodiversity might lead us to expect significant overlap in their images and descriptions. However, here their paths diverge. Farmers showed biodiversity through pictures that captured several life forms, mostly in landscapes (figure 10.3a). Their discussions were holistically oriented, encompassing plants, animals, and humans. The illustrations and explanations of NGO employees, on the other hand, concentrated on the species level (figure 10.3b). They showed pictures of farmers working potato fields and discussed the significance of human management of native crops, particularly the often neglected role of women. Their pictures further showed mountain plant communities, as well as gardens and street scenes. Garden photos portrayed potted native medicinal plants and plants of tropical origin. All of the scientists used photographs of potato varieties to illustrate the concept of biodiversity (figure 10.3c). They oriented their discussions to the role of native potato diversity and farmers' knowledge of it, showing scenes of farmers together with their varieties of potatoes. Differing from the NGO workers, their discussions were more detail oriented and focused on the varietal level.

Thus, despite the seemingly universal agreement on how to define the essence of biodiversity, the participants' representations varied distinctly in terms of scale: at the landscape (or ecosystem) level for farmers, the species level for NGO employees, and the intraspecific level for scientists. This differentiation might be tied to where they are coming from: farmers firmly rooted in a holistic Quechua cosmovision; scientists trained for detailed-level analysis and, particularly for CIP scientists, to focus on a mandated commodity—the potato; and NGO workers located in between, having been brought up close to the surrounding rural landscapes and lifescapes but at the same time exposed to larger chunks of other influences through formal education and job training.

Reflecting on Rights

Nobody can say that one doesn't have the right, no?
Farmer, Parque de la Papa

For me, a right is something that you, by being a person, already have. You don't have to do anything to achieve it, just by being [you have it].
NGO employee, ANDES

Rights, from the perspectives of farmers and NGO employees, are inalienable and intrinsic—they do not have to be earned. Scientists did not provide much explanation about how they conceptualize this term. Overall, the participants' discussions of rights covered much ground, in terms of both the range of examples provided and how rights are differently distributed between actors. A summary of the rights captured in the participants' images is presented in table 10.2, where the responses are sorted into categories that emerged from the data. Only one right appeared in the discussions of all groups: the right to food, or "food security" in scientists' terms. Further, it shows a high degree of similarity between the kinds of rights pointed to by farmers and by NGO employees. Farmers, however, provided more examples.

NGO employees' and farmers' illustrations covered a number of similar topics, such as cultural rights, rights to happiness and to a home, and work-related rights. In addition, farmers illustrated a number of rights not mentioned by the other groups: rights to participate in sports, to health, to keep animals, to clothing, and to education. They also pointed to rights that had to do with society, for example, the right to vote and the right to participate in meetings and workshops. Scientists devoted less attention to illustrating this term, but they discussed the right to information about the properties of native potato varieties, along with farmers' property rights to potatoes and their right to repatriation of na-

Table 10.2. Rights highlighted in participants' photographs

Participant group	Category								
	Food	Life	Home	Culture	Happiness	Work	Technology	Participation	Other
Farmers	Food	Live	House, with things and products	Typical clothing Customs-dance	Enjoy Be happy	Break from work	Green-house technology	Vote Workshop Road access Ride bike	Sports Health Animals Education Clothing
NGO employees	Food	Life	House, protection	Cultural identity Know culture and past	Enjoy Play Be protected, happy Family love	Work Decent work			
Scientists	Food security						Traditional technology		Information Property/ repatriation

tive potato varieties. In addition, one scientist pointed to farmers' right to technology. Interestingly, a farmer also illustrated this right, but the types of technology they talked about were different. The scientist showed a picture of hand hoes and emphasized the right to retain traditional technology, whereas the farmer pointed to greenhouses, an example of desirable modern technology that, from his perspective, they have a right to.

The examples brought up in the exercise also shed light on differential distribution of rights (see table 10.3). Many rights were discussed broadly as "people's rights" or "human rights," but rights were also attributed to specific groups based on age, occupation, or geography. One scientist brought up the question of whether farmers' rights should be adopted in the U.N. Universal Declaration of Human Rights. Farmers talked about rural people's rights vis-à-vis urban people's rights and emphasized that they should be equal. For example, one farmer talked about the right to play soccer as important since "a healthy body [builds] a healthy mind . . . because of that, it is not only those who live in the city who know how to play, also those who live in the countryside have the right to play, no?" Frequently, farmers specified children's rights, as when they pointed to rights to education, clothing, food, and happiness. Then there were community rights such as the right to road access. NGO workers talked on a more general plane about people's rights and also mentioned the U.N. declaration. They discussed children's rights to cultural identity. They attributed rights to plants as well, as one NGO employee's comment about a potato photograph illustrates: "There are things that I do not agree with, like in the CBD. From the Andean point of view, all plants have rights. A person has the right to the potato, and the potato has the right to live, to be looked after by the person, and protected."

The CBD establishes national sovereign rights over biodiversity to countries of origin (Article 3) and further demands that all those bearing rights to biodiversity be taken into account in the implementation of

Table 10.3. Rights pointed to in illustrations, by participant group

Participant group	Farmers' rights	People's rights	Children's rights	Rural/ urban people's rights	Community rights	Plants' rights
Farmers		X	X	X	X	
NGO employees		X	X			X
Scientists	X					

the CBD (Article 1). The International Treaty on Plant Genetic Resources also recognizes national sovereign rights to biodiversity (Article 10). Significantly, it further establishes farmers' rights, the implementation of which rests with national governments, through measures including "(a) protection of traditional knowledge relevant to plant genetic resources for food and agriculture; (b) the right to equitably participate in sharing benefits arising from the utilization of plant genetic resources for food and agriculture; and (c) the right to participate in making decisions, at the national level, on matters related to the conservation and sustainable use of plant genetic resources for food and agriculture" (Article 9, FAO 2001:12–13).

In addition to illuminating how the meanings of this concept vary between actors, the data from this case study may provide some additional understanding that can guide the implementation of the rights specified in the above cited treaties. For example, when considering "all rights over those resources" (CBD) and "farmers' rights" (International Treaty on Plant Genetic Resources), the nuances emerging from the participants' discussion are relevant. Through their images and discussions they revealed that even within a single farming community, a diversity of right holders exists, for whom different kinds of rights may have greater relative significance—rights of the community as a whole, rights of individuals, rights of children, and of the plants grown. Interestingly, several farmers highlighted the right to participation (e.g., through voting and workshops), much in line with some of the measures to implement farmers' rights outlined in the treaty. This indicates that realization of these measures might indeed be one promising way of recognizing farmers' contributions in fostering PGR. Including farmers in negotiation and decision making concerning the regulation of PGR, such as the designation of access and benefit sharing could be crucial in order to achieve effective regimes governing the use and the conservation of these resources.

Negotiating Benefits

> One cannot always get a benefit if one has not worked.
> *NGO employee, ANDES*

> It is something that is not necessary for life, but some people can obtain it.
> *NGO employee, ANDES*

NGO employees concurred that benefits are nonessential attributes achieved by some people in some contexts, and often, unlike rights, they have to be earned. Defining what a benefit is turned out to be a trickier task for farmers and scientists although they provided ample examples of what benefits could be. In participants' explanations, benefits differed from rights in that they were often linked with sources or preconditions. Moreover, in many cases, a benefit was itself presented as a further source of another benefit. Thus, taken together, the narratives emerging from the participants' discussions pointed to a network of benefits, to some extent going back to their sources. The different explanations are diagrammed below to give a clearer overview of the meanings conveyed in the pictures (see figures 10.4a, b, and c).

Scientists focused their explanations on benefits derived from PGR. Virtually all of their examples were benefits emanating from the activities of CIP (see figure 10.4a). They emphasized nonmaterial benefits to farmers such as empowerment, education, information, awareness, valuation, and recognition. These benefits were illustrated through photos from workshops and events co-organized by CIP. They also presented pictures of virus-free native potatoes and improved varieties of fava beans and potatoes, including varieties improved through participatory plant breeding. These were captured during formal handover ceremonies, and from fields where the varieties were grown. Scientists also pointed to benefits not directly generated by CIP. This included the story of the amancay flower (*Ismene amancaes*), an important cultural symbol in Lima's coastal region, that was threatened to extinction by development but recently received a protected status through a private initiative (Perich Landa 2006). According to a CIP scientist, this initiative is derived from and reinforces the benefit of information and awareness about conservation. Another scientist pointed to a scientific article he had prepared, which he considered a benefit to the government and the country. One photograph showed a farmer having benefited from seeds of lettuce and other vegetables from a North American friend that the farmer used to start his own local seed production.

NGO employees only provided a few benefit illustrations, but the pictures they took were quite varied in content (see figure 10.4b). Their images showed material benefits from harvests of corn and potatoes and monetary benefits from selling on the street in Cusco. The exclusivity or lack of access to some material benefits was emphasized in two illustrations. One photo showed a poor man carrying a heavy load past a store. The explanation was that the store contains goods that the hard-working man could probably not afford. A set of photos showed two houses, one just up the street from the other, but much more beautiful, according to the photographer. The two pictures illustrated the idea that benefits, unlike rights, are not for all; the benefit of goods from a store or of a beautiful home is achieved by only a few. Finally, one photograph depicted a man sitting in the sun, relaxing and enjoying the sunshine, a benefit that is there for all, so evident that it may not be appreciated.

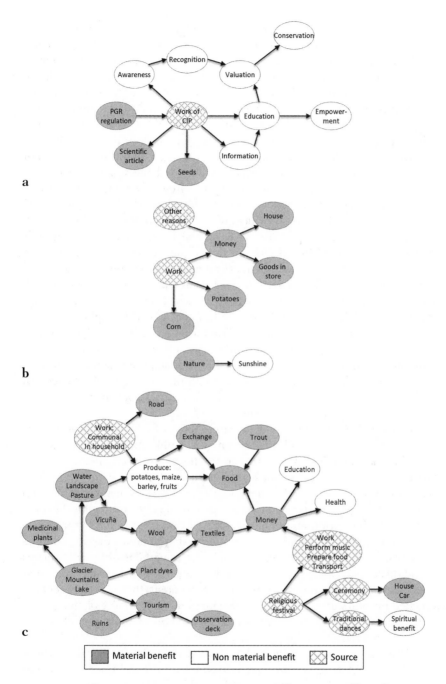

FIGURE 10.4. Diagrams summarizing participants' illustrations of benefits pointed to by (a) scientists, (b) NGO employees, and (c) farmers.

Farmers presented a diverse array of benefit illustrations (see figure 10.4c) that accrue to people in the surrounding rural landscape. Their pictures depicted agricultural products, the work and natural resources that go into their production, and the crucial roles these products play in daily life. They talked about communal work and the sharing of harvest. One farmer explained: "They harvest potatoes, and thus there are benefits to eat and sell." Another pointed out: "The potato is a very important food, right? In everything, at breakfast, at lunch, at dinner, just potatoes, nothing more, that way we are always benefiting [from this crop]." Other recurring themes in farmers' pictures included the weaving of textiles offered for sale to tourists, an important source of benefits in the form of income for many households. Another theme was the landscape and its constituents. Glaciers are providers of benefits in that they keep the landscape intact, nourish pastures, and are the source of water for animals and people. Mountains provide medicinal plants, whereas lakes and the entire landscape promise benefits through tourism. This was illustrated with photos from the construction of a *mirador* (a building from which tourists can enjoy the view), part of a recent ecotourism initiative supported by ANDES in the Parque de la Papa, and with a photograph of ruins in a regional town. Pictures from a religious festival depicted ceremonies in which people build miniature cars and houses in the hope of obtaining full-size versions in the future. Dance performances portrayed spiritual benefits; through the dance their burdens are lifted. The festival also provided ample economic benefits from food sales, horse and mule transport, and music performance.

Taking all groups' input into account, benefit is a complex concept that encompasses a number of tangible and intangible attributes. It should be noted that the only common thread to all groups' discussion of benefits was the potato—reflecting the central position of this food crop in Andean livelihoods—but they talked about different kinds of benefits in relation to it. Whereas scientists mostly used it to discuss intangible benefits such as information and empowerment, farmers and NGO representatives pointed to benefits in the form of food and money derived from its harvest. This difference runs through the remainder of their examples and discussions as well. With very few exceptions, the farmers and their advocates brought up material benefits, whereas the scientists concentrated on nonmaterial examples. Farmers more than any of the groups discussed economic benefits; indeed, nearly half of their examples were income related.

Fair and equitable benefit sharing is a recurring theme throughout treaty texts. Article 8(j) in the CBD establishes sharing of benefits arising from the utilization of indigenous and local knowledge, innovations, and practices. Article 15 stipulates the sharing of research results and benefits from commercial and other utilization of genetic resources, on mutually agreed terms, and this is further elaborated in Articles 16–21. The International Treaty on Plant Genetic Resources outlines a multilateral system for access and benefit sharing of the PGR belonging to a negotiated set of crops. Article 13 establishes "the exchange of information, access to and transfer of technology, capacity-building, and the sharing of the benefits arising from commercialization" (FAO 2001:20).

The benefit-sharing mechanisms in these treaties include both material and nonmaterial options. According to the data gathered, scientists' perceptions appear to be more in line with nonmonetary measures such as technology and information transfer and capacity building, whereas farmers' understanding of benefits appears to be more in line with direct economic compensation. As previously noted, a benefit identified by scientists was increased recognition of the value of biodiversity on part of the farmer. Still, a CIP scientist, placing himself in the latter's shoes, envisioned a farmer saying: "I know that this scientist tells me that it [crop diversity] has a great value. But I go to the market and they don't pay me." This shows that even if nonmonetary benefits dominate his own conceptualization, this scientist is aware that farmers interpret the concept differently. Since different stakeholders' conceptualizations are likely to be reflected in their expectations of benefits obtained through the treaties' mechanisms, increasing this kind of cross-actor awareness could be a productive strategy for clarifying demands and possibilities and facilitating negotiation and subsequent implementation.

"Benefits" and "rights" are, according to the interpretations of the present participants, interrelated concepts. What is a benefit to some may be perceived as a right to others. For instance, scientists considered workshops a benefit, but farmers saw them as a right. In some cases, the same participant gave different interpretations of the same object; a scientist explained that information about the properties of native potato varieties was a benefit in one photo but described it as a right in relation to another photo. An NGO employee talked about children's right to have fun in one photograph, and their enjoyment being a benefit in another. Another explained benefit as "a little like an extension of what a right is." A scientist illustrated the sliding scale between the two terms as he commented on crop commercialization: "It is not any more simply a benefit of being able to sell, but a right." Further discussion of how these concepts shade into each other is needed and could help in untangling expectations and accountabilities.

Conclusions

Through their illustrations and explanations thereof, Quechua farmers, ANDES NGO workers, and CIP scientists from Peru demonstrate diversity in their interpretations of key concepts employed in PGR regulation, in terms of relative salience as well as varying signification. Table 10.4 provides a summary of these situated meanings.

There was a marked difference between groups with regard to what concepts they chose to emphasize in their photographic work, indicating that the importance of the different concepts varied accordingly. This is a reminder that the importance of different key aspects of a regulatory text will vary between actors. Regarding the sites and situations used for photographic illustration, there was a division between scientists who chose only examples from their professional lives, and farmers and NGO workers who included images from both work and nonwork situations. This observation hints at differences between actors in terms of the role biodiversity and its regulation play in their lives. For farmers and NGO

Table 10.4. Summary of situated meanings of key concepts among participant groups

Dimension	Participant group		
	Farmers	NGO employees	Scientists
Concept priority	1. Benefit 2. Right 3. Biodiversity	1. Right 2. Biodiversity 3. Benefit	1. Benefit 2. Biodiversity 3. Right
Scenes chosen for illustrations	Personal and professional life	Personal and professional life	Professional life
Focus of biodiversity illustrations	Ecosystem level	Species level	Varietal level
Focus of right illustrations	People's rights Children's rights Rural vs. urban people's rights Community rights	People's rights Children's rights Plants' rights	Farmers' rights
Focus of benefit illustrations	Material and economic	Material	Nonmaterial

workers, these topics grip their whole lifeworld with no clear distinction between private and professional spheres, whereas for scientists it is apparently more linked to a distinct professional life.

The perceptions of the different groups regarding the concepts of biodiversity, right, and benefit overlapped to some extent but departed in several important aspects. When describing biodiversity all participants used similar words, neatly fitting the CBD's definition, suggesting a "bleeding" of concepts into the consciousness of different stakeholders with their involvement in a collaborative project, like the repatriation of native potatoes. Yet the focus of their cameras' lenses was pitched at different scales: farmers captured the entire landscape, and their holistic compositions included several life forms; NGO employees pointed to different species; and scientists emphasized intraspecific diversity. There was more overlap found in terms of how the groups illustrated rights, and especially farmers and NGO employees converged on many rights examples. Still, when delving into a more detailed level, differences appeared, as when the right to technology brought up images of traditional tools and techniques in the eyes of the scientist and modern inventions in those of the farmer. Regarding right-holder groups, scientists discussed farmers' rights, whereas the illustrations from farmers and NGO workers revealed a more nuanced picture of right holders within farming communities where different kinds of rights are attributed to the community as a whole, individuals, children, and plants. Lastly, in terms of benefits, there was a rather sharp distinction between scientists, who directed their discussions primarily toward intangible examples, and farmers and NGO employees, who presented images concentrated on material or monetary benefits.

Creating conventions that make sense to those who are affected is of course crucial, and regulatory texts have been criticized for their reliance on technical terms that are unintelligible for all but a small group of experts (Cleveland and Murray 1997). However, with the possible exception of biodiversity, the terms focused on here are not "artificial" constructs created during negotiations but part of an everyday vocabulary. "Biodiversity" originated in academic and policy circles in the 1980s but has quickly gained global popularity (Nazarea 2006), and none of the participants expressed confusion regarding the term. But, as we have seen, employing familiar words also presents a problem that cannot be discounted. The usage of concepts from everyday life such as rights and benefits may open up the possibility of having to negotiate a wide variety of connotations and expectations shaped in specific cultural contexts.

The present findings underline the necessity of action that can ensure mutual understanding of terms between different stakeholders that reaches beyond rhetoric, and certainly beyond authority or power.

The method employed in this study, photovoice, promises to be a useful instrument toward these ends. Since cameras are tools that can be managed by virtually everyone, the method helps to level an otherwise unequal playing field. It does not require a large time commitment on the part of participants or researchers in order to be effective and offers new insight via research analysis and focus group discussion. The results can be analyzed in a fairly rigorous manner at the end of the process, and communicated to a wider audience in written and visual forms. Yet the most important realizations might arise in the course of doing this exercise as participants reflect on what images to capture, carry out the photographic work, and share their images and reflections. If photographers representing different stakeholders are gathered, their pictures can serve as potent cross-cultural translators and, by stimulating interpretation and dialogue, can help achieve insights into each other's perspectives that are not easily obtained by oral description alone. As a research methodology, photovoice places the lens in the hands of the participants and thus empowers them.

Despite their different situations, the participants in this study were all citizens of the same country and belonged to institutions with a history of interaction. Among stakeholders on the global level, we may anticipate that images and interpretations will diverge on an even wider scale. Still, through candid, open communication, in which the viewpoints of different actors are given voice, commonalities and departures in each other's conceptualization can be discovered. As long as the assumption of homogeneity of interpretations of different groups or the supremacy of the understanding of one group over others are dismantled, these situated meanings can serve as starting points for finding common ground.

References

Agrawal, Arun. 1995. Dismantling the divide between indigenous and scientific knowledge. *Development and Change* 26(3):413–439.

ANDES (Asociación para la Naturaleza y Desarrollo Sostenible). n.d. Misión y visión. www.andes.org.pe/es/nosotros/misionvision.html. Accessed February 10, 2007.

ANDES, Parque de la Papa, and CIP. 2004. *Convenio para la repatriación, restauración y seguimiento de la agrobiodiversidad de papa nativa y los sistemas de conocimiento comunitarios asociados*, Asociación para la Naturaleza y el Desarrollo Sostenible, Asociación de Comunidades del Parque de la Papa, y Centro Internacional de la Papa, www.grain.org/brl_files/Convenio%20CIP .pdf, accessed February 10, 2007.

Bebbington, Anthony. 1997. New states, new NGOs?: Crises and transitions among rural development NGOs in the Andean region. *World Development* 25(11):1755–1765.

Berkes, Fikret. 1999. *Sacred Ecology: Traditional Ecological Knowledge and Resource Management*. Philadelphia: Taylor and Francis.

Berlin, Brent. 1992. *Ethnobiological Classification: Principles of Categorization of Plants and Animals in Traditional Societies*. Princeton, NJ: Princeton University Press.

Bourdieu, P. 1987. What makes a social class?: On the theoretical and practical existence of groups. *Berkeley Journal of Sociology: A Critical Review* 31:1–18.

Brush, Stephen B., and Doreen Stabinsky. 1996. *Valuing Local Knowledge: Indigenous People and Intellectual Property Rights*. Washington, DC: Island Press.

CGIAR (Consultative Group on International Agricultural Research). n.d.-a. History of the CGIAR, www.cgiar.org/who/history/index.html, accessed February 3, 2007.

———. n.d.-b. The CGIAR mission, www.cgiar.org/who/index.html, accessed February 3, 2007

CIP (Centro Internacional de la Papa). n.d.-a. Potato park, http://cipotato.org/gene bank/potato-park, accessed May 2, 2013.

———. n.d.-b. Mission statement, www.cipotato.org/org/mission.htm, accessed February 3, 2007.

Cleveland, David A., and Stephen C. Murray. 1997. The world's crop genetic resources and the rights of indigenous farmers. *Current Anthropology* 38: 477–515.

Ellen, Roy F., ed. 2007. *Modern Crises and Traditional Strategies: Local Ecological Knowledge in Island Southeast Asia*. New York: Berghahn Books.

Escobar, Arturo. 1998. Whose knowledge? Whose nature?: Biodiversity, conservation, and the political ecology of social movements. *Journal of Political Ecology* 5:53–82.

FAO. 1983. *International Undertaking on Plant Genetic Resources for Food and Agriculture*. Rome, Italy: Food and Agriculture Organization of the United Nations.

———. 1989. *Report of the Conference of FAO, Twenty-fifth Session, Rome, 11–29 November 1989*, Resolution 5/89.

———. 2001. *International Treaty on Plant Genetic Resources for Food and Agriculture*. Rome, Italy: Food and Agriculture Organization of the United Nations,

———. 2009. List of projects approved to be funded under the Benefit-Sharing Fund within the first project cycle, Food and Agriculture Organization of the United Nations, ftp://ftp.fao.org/ag/agp/planttreaty/funding/pro_list09_01_en .pdf.

Fowler, Cary, and Pat Roy Mooney. 1990. *Shattering: Food, Politics and the Loss of Genetic Diversity*. Tucson: University of Arizona Press.

Fowler, Cary, Melinda Smale, and Samy Gaiji. 2001. Unequal exchange?: Recent transfers of agricultural resources and their implications for developing countries. *Development Policy Review* 19(2):181–204.

Frankel, O. H., ed. 1973. *Survey of Crop Genetic Resources in Their Centres of Diversity.* Rome: Food and Agriculture Organization of the United Nations.

Haraway, Donna. 1988. Situated knowledges: The science question in feminism and the privilege of partial perspective. *Feminist Studies* 14(3):575–599.

Hawkes, J. G. 1985. *Plant Genetic Resources: The Impact of the International Agricultural Research Centers.* Washington, DC: World Bank.

Hayden, Cori. 2003. *When Nature Goes Public: The Making and Unmaking of Bioprospecting in Mexico.* Princeton, NJ: Princeton University Press.

Lansing, John Stephen. 2006. *Perfect Order: Recognizing Complexity in Bali.* Princeton Studies in Complexity. Princeton, NJ: Princeton University Press.

Lele, Uma J., and CGIAR. 2004. *The CGIAR at 31: An Independent Meta-evaluation of the Consultative Group on International Agricultural Research.* Rev. ed. Washington, DC: World Bank.

Marcus, George E., and Michael M. J. Fischer. 1986. *Anthropology as Cultural Critique: An Experimental Moment in the Human Sciences.* Chicago: University of Chicago Press.

McIntyre, Alice. 2003. Through the eyes of women: Photovoice and participatory research as tools for reimagining place. *Gender, Place and Culture* 10(1):47–66.

Nazarea, Virginia D., ed. 1999a. *Ethnoecology: Situated Knowledge/Located Lives.* Tucson: University of Arizona Press.

———. 1999b. Introduction. A view from a point: Ethnoecology as situated knowledge. In *Ethnoecology: Situated Knowledge/Located Lives*, edited by V. D. Nazarea. Tucson: University of Arizona Press.

———. 2006. Local knowledge and memory in biodiversity conservation. *Annual Review of Anthropology* 35(1):317–335.

New Scientist. 2005. Indians in Peru regain potato rights. January 22, 5.

Orlove, Benjamin S., and Stephen B. Brush. 1996. Anthropology and the conservation of biodiversity. *Annual Review of Anthropology* 25:329–352.

Perich Landa, Tatiana. 2006. Empresa emprende el rescate de la emblemática flor de amancay. *El Comercio*, August 9.

Petit, Michel, Wanda Collins, Cary Fowler, Carlos Correa, and Carl-Gustaf Thornström. 2001. *Why governments can't make policy—the case of plant genetic resources in the international arena.* Lima: International Potato Center.

Plucknett, Donald L. 1987. *Gene Banks and the World's Food.* Princeton, NJ: Princeton University Press.

Posey, Darrell Addison. 1994. Traditional resource rights: De facto self determination for indigenous peoples. In *Voices of the Earth: Indigenous Peoples, New Partnerships and the Rights to Self Determination*, edited by L. van der List, 217–235. Amsterdam: International Books.

Rosenthal, Joshua P. 2006. Politics, culture, and governance in the development of prior informed consent in indigenous communities. *Current Anthropology* 47(1):119–142.

Salazar, Milagros. 2008. Preserving the potato in its birthplace. *Inter Press Service News Agency Newsletter*, April 28.

Shiva, Vandana. 1993. *Monocultures of the Mind: Perspectives on Biodiversity and Biotechnology.* 1st Indian ed. Dehra Dun: Natraj.

———. 1997. *Biopiracy: The Plunder of Nature and Knowledge.* Boston: South End Press.

Sillitoe, Paul. 1998. The development of indigenous knowledge: A new applied anthropology. *Current Anthropology* 39(2):223–252.

Srinivasan, Chittur S. 2003. Exploring the feasibility of farmers' rights. *Development Policy Review* 21(4):419–447.

Suri, Sanjay. 2005. Potato capital of the world offers up new recipe. *Inter Press Service News Agency Newsletter*, January 18.

Tvedt, Terje. 2002. Development NGOs: Actors in a global civil society or in a new international system? *Voluntas: International Journal of Voluntary and Nonprofit Organizations* 13(4):363–375.

UNEP. 1993. *Convention of Biological Diversity.* Article 2. Nairobi: United Nations Environment Programme.

Van Rooy, Alison. 2000. Good news! You may be out of a job: Reflections on the past and future 50 years for Northern NGOs. *Development in Practice* 10(3/4):300–319.

Wang, Caroline, and M. Burris. 1997. Photovoice: Concept, methodology, and use for participatory needs assessment. *Health Education and Behavior* 24:369–387.

Warren, Samantha. 2005. Photography and voice in critical qualitative management research. *Accounting, Auditing and Accountability Journal* 18(6): 861–882.

Wynberg, Rachel, Roger Chennells, and Doris Schroeder, eds. 2009. *Indigenous Peoples, Consent and Benefit Sharing: Lessons from the San-Hoodia Case.* Dordrecht: Springer.

CHAPTER ELEVEN

Exile Landscapes of Nostalgia and Hope in the Cuban Diaspora

JENNA E. ANDREWS-SWANN

Immigrants are defined by their mobility. They are always and forever distinguishable from those born in the host country. On a day-to-day basis they negotiate ways around experiences and memories of homeland and experiences and realities in the host country.

Mariastella Pulvirenti, "Anchoring Mobile Subjectivities: Home, Identity, and Belonging among Italian Australian Migrants"

The history of Cuban migration to the United States is a perplexing, complex tale. It is a story that provides some interesting fodder for discussions of identity in exile, owing to the island's political past. Most notably, in the midst of negative repercussions of the 1959 Cuban Revolution, the U.S. travel and trade restrictions, and the subsequent dissolution of the Union of Soviet Socialist Republics, Cubans continue to leave the island in significant numbers. Scholars have often focused on the resultant Cuban diaspora (e.g., Behar and Suárez 2008; Duany 2005) and the effects of economically and politically motivated migration on the island and on places abroad. These migrants have settled in an array of locations around the world. However, U.S. policies toward Cuban exiles, historic political and economic agreements, ongoing cultural exchange between the two countries, and Cuba's geographical proximity to the United States have meant that a large majority of Cubans who have fled the island now consider the United States their adopted home. Indeed, the first large-scale migration from Cuba to the United States dates back to the Ten Years' War, which was waged in Cuba between 1868 and 1878.

Many Cuban immigrants who have arrived in the United States (and elsewhere throughout the diaspora) variously label themselves *exiliados* (exiles), *inmigrantes* (immigrants), or *refugiados* (refugees), depending in part on the circumstances of their departure and resettlement. The usage of each of these labels is undoubtedly political. "Exile" connotes a decision to leave Cuba and withstand a period of absence from the country; it suggests a strong stance against post-Revolutionary policies. The Cuban exile community in Miami is well known for its wealth and "success," but as Emily Skop (2001) notes, this community has been forced to incorporate new elements—in terms of both exiles' reasons for leaving and their social class—as subsequent waves of Cubans arrive in Miami. Of course, Cuban exiles in the United States are not unique to Castro's revolutionary government; they first arrived during the latter part of the nineteenth century in response to "Spanish colonial administration" (Pérez 1978:129).

Modern, contemporary migration the world over is marked by a new global connectivity that seems to seep into even the most remote and isolated of places. But Arjun Appadurai (1990:3) notes that "if 'a' global cultural system is emerging, it is filled with ironies and resistances." Diaspora is evidence of such resistance. That diasporas continue to exist amidst the forces of globalization is testament to the significance of maintaining an identity created within a particular spatial and historical context: the homeland. While akin to theories of transnationalism and border straddling, diasporas "usually pre-suppose longer distances and a separation more like exile: a constitutive taboo on return, or its postponement to a remote future. Diasporas also connect multiple communities of a dispersed population. Systematic border crossings may be part of this interconnection, but multi-locale diaspora cultures are not necessarily defined by a specific geopolitical boundary" (Clifford 1994:304). Members of a diaspora routinely share memories, longing, and nostalgia that may focus on a real or imagined homeland (Clifford 1994).

The kinds of landscapes that are created in diaspora represent collections of places, created through sensory experiences and characterized by their ability to gather and hold memory (Casey 1998; Abercrombie 1998; see also chapter 4, this volume). Charles Tilley notes that "geographical experience begins in places, reaches out to others through spaces, and creates landscapes or regions for human existence" (qtd. in Escobar 2001:15). A place is not simply marked by a set of coordinates; rather, it is imbued with specific meaning for the people that occupy it and thus is a potential source of identity and resistance (Basso 1996; Casey 1997). Places are not bounded areas, but open, with porous boundaries that allow a place to intermingle with its surroundings so that places and their identifying features are constantly reconstituted (Escobar 2001; Massey 1994). The contemporary prevalence of migration and global communication has challenged our traditional attachment to place and the landscape: "For some [scholars], placelessness has become the essential feature of the modern condition" (Escobar 2001:140).

A sense of place, then, based on real or imagined interaction with the landscape and all it encompasses, reflects attachment to a space and its accompanying smells, tastes, or feeling. Marginalized peoples, such as exiles, often express a remembered sense of place in opposition to mainstream society (Gordillo 2004; Dusselier 2002; Feld and Basso 1996). Cultural memory is embedded in the landscape, continually reconstituted in and through specific places (Connerton 1989; Rappaport 1998; Rigney 2005). Landscapes like the kitchen or the garden represent important sites of cultural construction, which can become more deliberate or palpable in the face of distressing events such as the recent food shortages in Cuba or the process of migration itself (Christie 2004; McDowell 2004; Sutton 1998). Thus, place and displacement become interrelated states of being, wherein (re)creating place becomes a site of power and resistance for those who have been displaced (see chapters 1 and 6, this volume).

This chapter addresses the volume's theme of conservation of biodiversity hinging on resistance and hope by considering some of the ways that nostalgia works to create a sense of place and ethnic identity in the face of homogenizing state policies. I draw on the results of fieldwork conducted in two sites that are intended, at least in part, to represent some of the diversity of Cuban exile experiences. In each city, one significantly more urban than the other, exiles "re-member"—revise, remake, and rethink—their surroundings into something that reflects their memories and desires. Yet, because of the particularities of place, the mechanisms and dimensions of recreation are different in each site. In these transplanted and reconstructed places, the seeds of resistance and hope manifest seeds of agency, memory, and strength that sprout into landscapes that become domesticated and familiar enough to call one's own.

Transnational Nostalgia and Memory

Nostalgia is a painful homesickness that generates desire . . . nostalgia sets in motion a dialectic of closeness and distantiation.

Kathleen Stewart, "Nostalgia—A Polemic"

Memory and nostalgia spur so much interest because these concepts are simultaneously academically challenging and personally relatable (Berliner 2005). But they are complex and messy ideas. "More and more diasporic groups have memories whose archaeology is fractured. . . . Even for apparently well-settled diasporic groups, the macro-politics of reproduction translates into the micro-politics of memory, among friends, relatives and generations" (Appadurai and Breckenridge 1988, qtd. in Vertovec 1997:9). In his discussion of one Cuban American landmark, the Our Lady of Charity shrine in Miami, Thomas Tweed (1997:148) notes that collective ritual around the Virgin has important implications for Cuban exiles' shared national identity: common ritual bridges the diaspora with Cubans on the island, "creating an imagined moral community and generating feelings of nostalgia, hopefulness, and commonality." That is, at some level, and despite disagreements over the meanings attributed to a particular symbol, the Cuban community abroad shares some sense of collective identity that shapes the way they re-member the past.

Nostalgia has been described as a particular longing for something that once was, or something that may have been (Seremetakis 1996), often evoked by sensory triggers such as smells or tastes (Sutton 2001). This longing may also apply to something that one never directly, personally experienced (Appadurai 1996). It is an interesting concept for analyzing the roles that memory plays in the context of migration because nostalgia is not limited to firsthand experiences. Rather, it is associated with a sense of longing for the past, whether one experienced it or not. The concept of nostalgia thus highlights the flexible and confounding nature of memory. For example, contemporary Afra-Cuban writers live and work in a time when slavery is no longer legally sanctioned, but many continue to conjure images of slavery in order to situate their contemporary existence (e.g., Nancy Morejón, Excilia Saldaña). Cuban American artists often present images of barely seaworthy boats jostled by waves, though many made the trip in airplanes or were born in the United States. But this generative quality of nostalgia also has the effect of creating a broad, rather diverse community around shared memories, potentially contributing to "diasporic nationalism"—a sense of nationalism abroad that bridges the gap between a remembered homeland and the adopted one (Tweed 1997).

Nostalgia and memory often fill in the gaps left by infrequent or impersonal contact with "home," resulting in a diverse array of individualized landscapes made by people who have experienced migration and exile in distinct ways. In addition, the case of Cuban migration to the United States is also informed by specific international policies and the connections people maintain with the island. Attempts to ease feelings of loss are often the basis for establishing and maintaining transnational networks—bridges between Cuba and the United States—which in turn inform the ways that exiles and émigrés experience a sense of national identity abroad. By focusing on the "interconnected realm of cross-border relationships," we can better understand the processes of migration and integration as they interact with the structure of the nation-state (and our own role in explaining these processes) (Wimmer and Glick-Schiller 2003:586). Alejandro Portes (1997) notes that immigration research has a history of focusing on individual cases; he argues that the field would benefit now from distilling this work into a broader theory of migration, a perspective generated from the observation that contemporary migrants are becoming increasingly, and nearly uniformly, transnational (Glick-Schiller et al. 1992).

This transnational character of contemporary migration recognizes that the geographical proximity of homeland and host land has become less important to the migration experience in general because of increasingly fast, accessible, and global connections. National borders are no longer indicative of a homogenous people (Smith and Guarnizo 1998). Instead, they are figurative lines to be crossed in an effort to, say, earn more income or gain new freedoms. Border crossing does not rule out returning home or traveling on to another place; in fact, transnationalism facilitates this back-and-forth movement of people and ideas through porous boundaries. Indeed, borders are crossed even in such everyday activities as watching television or consuming "ethnic" foods. The vast literature on transnationalism is one indication "that the nation-state container view of society does not capture, adequately or automatically, the complex interconnectedness of contemporary reality" (Levitt and Glick-Schiller 2004:6).

Transnational social networks are often at the heart of immigrant "enclaves" that tend to involve economic interactions between people of the same cultural or ethnic background. Enclaves represent a collective immigrant or ethnic identity in opposition to a mainstream society that offers little in the way of social or economic support. Not only are enclaves considered sources of economic resources, but various authors (e.g., Mahler 1995; Nagel 2002) note that they may provide a forum of sorts for immigrants to voice collective political or social concerns as well. Miami is often used as a case study to discuss the phenomenon of urban ethnic enclaves (Cobas and Duany 1997; Logan, Alba, and McNulty 1994; Portes 1987). "Groups such as Cubans in Miami have adapted . . . through an 'ethnic enclave', that is, a concentration of interrelated businesses that occupy a distinct territory and serve primarily the in-group" (Cobas and Duany 1997:2). Others note that there are important limits to the strength of the enclave, notably that immigrant workers who are successful wage earners tend to integrate into the mainstream prior to immigrant entrepreneurs (Sanders and Nee 1987).

But to make matters more confusing, assimilation and transnationality are "neither incompatible nor binary opposites" (Levitt and Glick-Schiller 2004). That is, transnational connections do not necessarily indicate that we are moving toward one big, blended global culture. Rather than conceptualize transnationality and diaspora as global economic forces, the fuzzier perspectives of diaspora and place making remind scholars to recognize the role of human agency in (re)creating culture across different spaces (Ong 1999). This condition is reminiscent of Steven Vertovec's "diaspora as a type of consciousness," wherein a rather paradoxical set of experiences with the homeland determine one's sense of identity (Vertovec 1997). This consciousness is simultaneously informed by negative exclusionary experiences, positive experiences of a shared cultural heritage, and a desire to connect oneself with others. Tweed echoes this role of the diaspora in his discussion of a "sense of geopiety, or an attachment to the natal landscape that, for Cuban Americans, reflects the utopia of memory and desire, not the dystopia of the contemporary socialist state" (Tweed 1997:132).

So memory involves a degree of performance and ritual; the past is remembered, "conveyed and sustained" by ritual performances, and these may occur more strongly in and through familiar landscapes than foreign ones (Connerton 1989:40). Thus, the impetus for (re)creating a landscape based on ideas of home. It is necessary to address this dialectic nature of contemporary, often transnational, migration and nostalgia, and acknowledge that it cannot be understood without adequate attention to the ways that connections are established and experienced across oceans and national borders by individuals, communities, or government entities as they each dwell in and re-member familiar landscapes.

As Lucía Suárez (2006) notes, there exists an important tension between memory and forgetting. In light of the violent, tumultuous history of the Caribbean, affirmative, hopeful memory is necessarily thrust into stark contrast with the impossibility of bringing back the dead, in both a literal and a figurative sense. The relationship between memory and forgetting, however, is not dichotomous; rather, the two exist as a continuum moderated by people who re-member, reclaim, and reimagine history. Re-membering, whether through food, language, music, or monuments, recontextualizes the past; that is, the past—reshaped and reclaimed—becomes rooted in the present as it is integrated into a living, breathing presence once more.

Food-centered nostalgia is a common theme in migration literature (Choo 2004; Armstrong 1999), and food often serves as a gateway for immigrants to engage with collective representations of homeland or national identity (Ray 2004; Mankekar 2002). Food is a particularly important element by virtue of its everyday presence, and sensory memories related to food become especially apparent when particular foods are no longer available (Seremetakis 1996). Embodied memory and emotional experiences play a significant role in the way sensory images are recreated and *placed* (Sutton 2001); the homegarden and the kitchen, for instance, are often considered resilient places of remembrance (Christensen 2001; Nazarea 2006).

These forces contribute to the creation of immigrant landscapes that represent a confluence of nostalgia, cultural tradition, and active place making. Sonia Graham and John Connell (2006) found that among Greek and Vietnamese immigrants in Sydney, Australia, the environment created by a garden, along with the garden produce, helped immigrant gardeners emphasize and maintain cultural relationships and social networks, provide a space for nostalgia, and give immigrant gardeners a sense of ownership and control. Gardens and yards become physical manifestations of memory that embody family and community traditions (Graham and Connell 2006). Immigrant gardens may also serve as sources of income, as locals and other immigrants take advantage of exotic and familiar produce grown by members of their community. For example, Southeast Asian immigrants in Homestead, Florida, have created a niche for themselves by focusing on growing specialty Southeast Asian herbs, fruits, and vegetables (Imbruce 2007). In effect, gardens such as these represent a *trans situ* form of agrobiodiversity conservation through use since seeds and other plant matter are often brought to the United States from a multitude of homelands (Nazarea 2005; see also chapter 12, this volume).

Within immigrant communities, common language, food preferences, religion, and various other shared cultural traditions reinforce a sense of belonging, all of which may reduce the effects of acculturation (Graham and Connell 2006; Airriess and Clawson 1994). For immigrants—and, it may be argued, for other marginalized groups—the traditions maintained through social networks may in turn contribute to a shared nostalgic representation of an idealized homeland or history (Dawdy 2002).

Remembrance in the Cuban Diaspora

> You're "cubanglo," a word that has the advantage of imprecision, since one can't tell where the "Cuban" ends and the "Anglo" begins. Having two cultures, you belong wholly to neither. You are both, you are neither: *cuba-no/ america-no*.
>
> *Gustavo Firmat, Life on the Hyphen*

By re-membering, nostalgia, memory, and a sense of place are woven into lived landscapes, showcasing the ways that members of the Cuban diaspora in the United States have creatively circumvented the laws that minimize their ability to interact with the homeland by bringing the "homeland" with them. This is reflected in the diaspora experience in myriad ways, dependent, in large part, on the context(s) of exile and resettlement. The discussion that follows features some of these experiences in Miami, Florida, and in Moultrie, Georgia.

Miami, Florida: Havana, USA

Miami has come to be synonymous with Cuban influences. Language, music, cuisine, architecture, and even local political persuasions in Miami have all developed over the past century or two in tandem with South Florida's relationship with Cuba and Cuban people. Cubans continue to exert a strong influence over the region in part because of their economic and political success, but also because of their sheer numbers relative to other minority groups. And because so-called new Cubans continue to arrive and Cuban Americans periodically return to Cuba, the region's connectedness to Cuba remains vital. Many Miami institutions feature this connection prominently. The extensive Cuban Heritage Collection at the University of Miami; *El Nuevo Herald*, the Spanish language counterpart to the *Miami Herald* daily newspaper; and the Archdiocese of Miami are just a few examples of the formalized connections to Cubans and Cuban culture that are maintained and made publicly accessible to Miamians. Scores of popular and academic publications have presented accounts of Cuban (and Cuban American) lives lived in Miami, or "Havana USA" (Garcia 1996).[1] Many of these works also outline the influence of the political agenda in line with the majority opinion of first-generation Cuban refugees in Miami: Castro's government must be ousted lest it commit more abuses. More tempered perspectives toward the revolutionary government have developed among those Cubans that either grew up in the United States or were born here, but a sense of loss pervades in much of this segment of the community as well. In general, the conservative perspective so popular in Miami has promoted a nostalgic longing for a place that has been ravaged and, according to many, ruined by communism. Monuments stand as testament to this sense of loss and longing to such an extent that it has become an integral part of Cuban landscapes in Miami.

Little Havana remains the cultural center of Cuban Miami. Each month, the neighborhood is host to Viernes Cultural, or Cultural Friday, an evening event that combines Cuban music, food, crafts, fine art, and cigars. The gathering attracts Cubans from all over the city. The store-fronts in Little Havana suggest the area's Cuban heritage as well. *Guaya-beras* (the linen dress shirt popular among Cuban men), social clubs, restaurants and cafeterias, *cafecito* stands, Cuban grocers and bakeries, and cigar shops are among the offerings on Calle Ocho, the main thoroughfare in Little Havana (see figure 11.1). Here, busy Domino Park is designated for use by residents over sixty years of age, and the tables are usually manned by exiles puffing on large cigars as they enthusiastically slap dominoes onto the boards. Most signage on Calle Ocho is in Spanish, and proprietors rarely speak English to customers. Long-time Cuban Miami residents have tended to move away from Little Havana and into more up-scale surrounding neighborhoods in Coconut Grove or even as far as suburbs in Hialeah and Kendall, but the shops, restaurants, and cultural events that take place in Little Havana always bring them back. The overall effect is one of in-betweenness—a self-reflexive, nostalgic sense of both relief and loss pervades this landscape. There is joy in the sumptuous tastes and sounds, but it is tempered by thoughts of what was left behind.

One particular event that exemplifies creative forms of resistance that perpetuate remembrance in the face of loss is CubaNostalgia: "A three-day event showcasing Cuban life, customs and heritage through exhibits, vendors selling Cuban-themed items, and memorabilia, art galleries and

FIGURE 11.1. Calle Ocho, Little Havana, Miami, Florida. Photo by Jenna E. Andrews-Swann.

artists" (CubaNostalgia 2009 event pamphlet). The original event took place in 1998 and was the brainchild of two prominent Cuban American businessmen who began the "showcase" in part because of their own personal interest in Cuban memorabilia (see figure 11.2).

The display of Cuban culture and the overall atmosphere of CubaNostalgia live up to the event's name. Displays and exhibits hearken back to a time when life in Cuba was "freer"—especially for those who benefited from the U.S. political and economic control of Cuba. But one Cuban American man in attendance, a photographer for *El Nuevo Herald* in his late twenties, noted in response to the event that "Miami has lots of layers . . . what you see here is not all of Cubans." The image of Cuba at CubaNostalgia is largely a glossed and homogeneous representation. It is a version common to CubaNostalgia's wealthy, conservative, white planners and sponsors, many of whom come from similarly wealthy families that were robbed of their livelihood in Cuba in the early 1960s. This community is inclined to remember particular aspects of prerevolutionary Cuba, despite the problems of that time, as home, and silence others. Owing to the power they wield in South Florida, other public representations of Cuba and Cuban heritage in Miami often mimic the kind of

FIGURE 11.2. CubaNostalgia entrance, Miami Fair Expo Center, 2009. Photo by Jenna E. Andrews-Swann.

Cubanness presented at CubaNostalgia. But this version of Cuban land-scape, with its 1950s-era decadence and inattention to more recent events (and arrivals), is not shared by all members of the Cuban exile community.

Generally speaking, the waves of emigration from Cuba since the 1959 revolution have become poorer and more racially diverse over time, reflecting the intersection of race and class politics in Cuba that prompted many wealthy white business owners to jump ship soon after the revolu-tion. Those of less means had fewer options and have remained in Cuba longer. The popular (though misconceived) representation of *Marielitos*, the nickname given to those who took part in the 1980 Mariel exodus from Cuba to the United States, as criminals or degenerates has also stuck in the minds of people in the United States, prompting further discrimination against recent arrivals (Croucher 1997). The following quote comes from an interview with Calle Ocho restaurateur Andrés,[2] who came to Miami on a private boat during the Mariel boatlift:

> I have a lot of nostalgia about Cuba. I love the Cuban music, the Cuban people. If you notice, my English is not that good. It's because I've always been in the Cuban neighborhood with the old people that play domino in the park. I'm always talking to them, and trying to read a lot about my place. I see my Central American workers here or other places . . . they take a vacation and they go to Colombia or El Salvador, and we can't. We have to ask permission to go to our home place. It's very sad. There's a lot of sadness among the Cuban people that can't go visit their relatives there that died. It's very sad. I want to take my chil-dren to the family farm with the horses. It's very sad.

This is a sentiment that is shared across much of Miami's diverse Cuban community, longing for what was or what could have been in Cuba, and for family or friends or memories left behind. Most people agree that Cuba is not the same now as it once was. The mainstream nostalgia that pervades expressions of Cuban identity in Miami is reminiscent of the widely held notion that Miami now represents the "real Cuba" in con-trast to the island that has been ravaged by the effects of a half-century of communist leadership. Resistance to that ruined version of home is reason enough to re-member things a little differently.

Moultrie, Georgia: The City of Southern Living

Nearly five hundred miles north of Miami, Cuba is re-membered quite differently by many in the Cuban community who now call Moultrie home. While overt, public nostalgia and political discontent are often center stage in Miami, food and gardens play a key—though quiet—role in the relatively new Cuban community in Moultrie as exiles there remember the island. Moultrie is not particularly known for its ethnic diversity or multiculturalism, but within the last few decades its Hispanic population has skyrocketed. The Cuban community has grown to nearly two hundred households, most of which have come to the United States within just the past twenty years (U.S. Census 2000). Moultrie is the county seat of rural Colquitt County; the region's agricultural economy gained momentum in the 1960s, and by the end of the decade the county was the number two producer of tobacco and cotton in the state of Georgia, bringing in close to $25 million (Colquitt County Community Assessment 2007). Contemporary agricultural endeavors attract migrant workers nearly year-round.

A unique hallmark of this segment of the Cuban diaspora is its religious affiliation: nearly all of Moultrie's Cuban families are practicing Jehovah's Witnesses. Several of the people interviewed reported having been imprisoned or otherwise oppressed in Cuba for actively practicing their religion, a situation that continues to prompt emigration from Cuba. Applying for asylum or refugee status to come to the United States, however, means renouncing the Cuban government, which complicates the prospect of returning one day to visit family left behind. These experiences in Cuba make many Cubans in Moultrie wary of outsiders, authority figures, and even inquisitive researchers. Interviewees often agreed to participate only on the condition that no questions be asked about politics or religion, and even then, many declined to be tape-recorded or photographed.

In this context, *cafecitos* (those ubiquitous thimblefuls of strong, sugary Cuban-style espresso) are brewed and enjoyed not at storefront windows as in Miami or Havana but in private kitchens in Moultrie households. The symbolic *cafetera* (stove-top coffee maker) and tiny porcelain espresso cups are a staple in these kitchens, along with a *tostonera* (wooden or plastic plantain press) and a deep fryer for making *tostones*. And although royal palm and ceiba trees, both classic Cuban landmarks, are a rare sight in Moultrie, some members of the exile community recall a familiar landscape in small garden spaces that blossom with a creative mix of both remembered and local edible and ornamental plants. Indeed, rural Moultrie itself is often a more familiar landscape for these exiles than Miami or other popular Cuban destinations in the United States, since a large majority has come not from urban Havana but the rural towns of eastern Cuba.

Hernán and Juana live on the outskirts of Moultrie in a trailer on a rural lot, and they have been in Moultrie for about twelve years. They are both retired, and recently Hernán has developed some health troubles that keep him close to home. He has not visited Cuba since leaving: "There are problems with religion and visas . . . it is my homeland, and unfortunately I had to migrate to the U.S. But it was a good idea to migrate." In their yard, Juana has planted several rose bushes along with mint, a tiny lime tree, and a few annual flowering plants that she grew in Cuba. Juana and her husband wanted this place in the country because it reminded them of "home." They have a vegetable garden along one side of the house with rows of *calabasa, pepino, ajo, boniato, yuca,* and *malanga* (squash, peppers, garlic, yucca, and taro). These are some of the traditional staple foods in Cuban cuisine, but others they tried to grow would not tolerate the cooler climate in Moultrie. Juana and Hernán cultivate these plants largely because the taste of their own produce is dramatically different when they are bought from a grocery store. Many other interviewees agreed: *really, truly* fresh meat and produce are much more difficult to find in the United States than they were in Cuba—even with food shortages on the island. The taste of packaged foods is considered inferior to fresh, unadulterated versions that were significantly more common (and affordable) in Cuba.

Olga is an energetic, cheery woman with light brown curly hair who lives in a new house in Moultrie along with her husband, Antonio, and their two young children. She is a homemaker and spends much of her time attending to her son, who is developmentally disabled. Olga grew up in a place in Cuba much like Moultrie, a small town surrounded by farms. But she noted that "it is very different because here, no one visits—everyone lives behind closed doors. No one is going to visit anyone. And in Cuba, no. In Cuba people visit you, doors are always open. It is more familiar." But she enjoys that she can have her own house in Moultrie, reflecting the trend there for Cuban residents to increasingly shift from low-income housing or rental properties to owning a home. This is a shift that for many is rife with political undertones in contrast to Cuban policies disallowing private ownership, which have largely remained in place for the past sixty years.

While feeling a bit isolated because no other members of Moultrie's Cuban community live nearby in her suburban neighborhood, she acknowledges that her neighbors are nice. She drew attention to her family's friendly interaction with people in Moultrie outside the Cuban community, especially for her young daughter, noting of her family, "no somos racistas" (we're not racist). While this emphasis may have been provided for my benefit, as Jehovah's Witnesses traditionally emphasize in-group relationships, it also may simply be a response to living as part of an ethnic minority group in a diverse community. The language barrier is difficult in Moultrie,[3] more difficult than it was in Tampa, Florida, where Olga and her family lived before settling in Moultrie—she knows just "*unas palabritas"* (a few little words). This makes doctor's appointments and meetings with her children's teachers challenging. For Olga, "the food here is better in the sense that there is more variety. But the tastes, how can I explain, in terms of flavor, there are things in Cuba that are better, like fruit—it's better." She and her family kept hens and goats in Cuba, primarily for food. In Moultrie she can recreate her family's tradition of *criando* (raising) animals in the backyard (there were several chickens clucking away as she spoke); she also plans to plant a vegetable garden like the one she remembers from her parents' house in Cuba, filled with whatever familiar edible plants she can get her hands on. Moultrie, unlike Tampa, gives her space for these familiar landscapes.

Homegardens are a source of re-membered familiarity in Moultrie, but because certain longed-for tastes and smells that cannot be cultivated are otherwise difficult to find, "orders" for such items are periodically placed with friends or family traveling south to Miami, where resources are more abundant and flavors more accessible. In general, though, and despite the difference in quality and freshness of some ingredients, Cuban food traditions continue to flourish in Moultrie. One resident noted that children born in the United States often will lose a taste for Cuban food, opting instead for the hamburgers or pizza and French fries they get at school. But she continues cooking mostly Cuban food in her home because she sees food as an important link to her children's heritage. Food is also a relatively easy way to express and remember a sense of Cuba, especially for those wary of being noticed; it is private and inoffensive, and it does not tend to draw attention from members of the majority. On one occasion, though, some non-Cuban neighbors called the local police after seeing what they thought was a dog being roasted in a Cuban family's backyard—it turned out to be a pig. The smells and textures of cooking and eating evoke happy memories of Cuba and create a sense of community among those that share in them together.

Idalma emphasizes food as a means to (re)create her Cuban identity in Moultrie. She came to Moultrie in 2004 with her husband and two daughters to avoid religious persecution. The family shares a tidy house near Moultrie High School. When Idalma learned of my interest in the way her community upholds particular Cuban cultural elements, she insisted that I come to her home. When I arrived, she was in the early stages of preparing a sumptuous feast. On the counter in her neat kitchen was the quintessential set of cafecito cups and saucers, orange and white with tiny flowers. Into a food processor went handfuls of onions, garlic, and cilantro. Idalma then skinned and quartered a chicken from Sanderson Farms, one of her workplaces, which went into a large and obviously well-used pressure cooker along with tomato sauce and paste and the fragrant mixture from the food processor. While the chicken began to cook, she toasted whole cumin seeds in a small pot, adding a little oil and a pinch of salt, which all went into the pressure cooker as well, followed by several cups of rice.

As Idalma fussed over the electric deep fryer, she spoke of the role of cooking and its importance to Cuban women: "Cocinar es integral . . . completer una mujer cubana" (cooking is key . . . it completes a Cuban woman). Idalma does not tend to use recipes for "las comidas tradicionales" (traditional foods) since she has simply grown up eating and cooking most of these dishes. She prefers the fresher taste of foods that she remembers from Cuba, but like other interviewees, she noted the much wider availability of a broad range of food items in the United States. And she so prefers the taste of saltwater fish that she planned a trip to Florida to purchase fish to prepare with *congrí* (a traditional Cuban dish of red beans and rice).

When the deep fryer was nearly ready, Idalma began squashing green plantain medallions in her plastic tostonera, purchased from a Winn-Dixie supermarket in Florida, and dropping them into the hot oil. As they cooked, she combined green beans and diced tomatoes with a squeeze of lime juice and a little salt, apologizing for not having her usual assortment of fresh vegetables on hand (beets, carrots, cucumber) to make a truly Cuban salad. She cooks similar meals for her family most evenings and happily reported that her daughters still love Cuban food, despite the pizza and burgers they are fed at school.

Meals like this one are common at religiously themed gatherings that serve as a common reason for exiles in Moultrie to come together, especially when people may be scattered throughout several neighborhoods. The Cuban Witness community is a tight-knit group, with most of those interviewed reporting that they and their children attend up to three Jehovah's Witness meetings weekly. These meetings are conducted in Spanish with fellow Cuban immigrants, both in the Kingdom Hall and in private homes. Other social gatherings revolve around the Witness community in Moultrie as well, because, according to one community member, "you can just tell" about those that are not Witnesses—prohibitions and a strict moral code distinguish members and nonmembers. Witnesses form a "millennial sect which emphasizes the approaching final judgment. . . . Witnesses are expected to be involved with the secular world as little as possible" (Alston and Aguirre 1970:64–65). In general, gatherings are limited to the Cuban community, although large affairs like baby showers may also include Mexican Witnesses (and white scholars) as guests. Most of these social affairs feature Cuban foods like roasted pork, rice and beans, *batidos* (tropical fruit milkshakes), or flan. Such familiar textures and flavors, combined with familiar religious perspectives and familiar language, privately and joyfully recall and re-member a shared identity. This kind of nostalgia carefully avoids negative memories associated with ideas of "home."

Conclusions

> The deterritorialization of identities . . . can result in a sense of local isolation, estrangement, and exclusion. Another consequence may also be a compensatory and nostalgic "place building" and an attempt to affirm the meanings and memories that are perceived to be threatened and soon to be lost.
>
> *Fiona Allon, "Nostalgia Unbound: Illegibility and the*
> *Synthetic Excess of Place"*

It could also be said that the *reterritorialization* of identity can be an effective mechanism for maintaining those meanings and memories associated with the homeland while resisting acculturation or the disappearance of important cultural attributes in a host country. Commenting on the sometimes freeing nature of "the between," Paul Stoller explains:

> Living between things can have several existential repercussions. It can simultaneously pull us in two separate directions so that, in the end, to quote a rural incantation of the Songhay people of Niger, "you don't know your front side from your back side." This state usually leads to indecision, confusion, and lethargy. . . . If, however, we find a way to draw strength from both sides of the between and breathe in the creative air of indeterminacy, we can find ourselves in a space of enormous growth, a space of power and creativity. (2008:4)

In both Moultrie and Miami, it is clear that the desire for collective or communal and individual expressions of Cubanness has drawn from "both sides of the between." The context(s) of exile and resettlement shape the ways that aspects of landscapes will be re-membered or forgotten to reflect a common diaspora identity.

In Moultrie, with its small exile community in a traditionally homogeneous town, the Cuban population has overwhelmingly turned inward for support and a sense of belonging in the context of exile. Nonpublic places, like kitchens, gardens, and the Hall, are the settings—or *milieu*, to echo arguments presented in chapters 1 and 2 of this volume—for family and community interaction, unlike the large festivals and public parks and monuments that Miami Cubans incorporate into their remembrances, which might be thought of as more powerful, somewhat standardized *sites* of memory. The style of nostalgia that characterizes Miami's version of Cuban identity is largely absent in Moultrie for a host of reasons.

First, Moultrie does not have the same intimate, historical relationship with Cuba that Miami does. Second, while much of the mainstream Cuban nostalgia presented in Miami comes from those who left the island (especially from Havana) in the very early years of the revolution, most Cuban residents in Moultrie lived through some of the leanest years of postrevolutionary Cuba, and many have suffered religious persecution to boot. Their image of Cuba as a nation and an identity is arguably shaped more by personal hardship than memories of 1950s opulence, and its expression differs in kind. That is, careful attention is paid to particular elements of Cuban culture and Cuban identity, which have little to do with the revolutionary government or its policies. The community's emphasis on food, language, and religion exemplifies this departure from the more overtly politicized public landscapes in Miami. Traditions are valued and maintained in Moultrie, but only insofar as they are divorced from the contemporary Cuban political climate.

In our global context of transnationalism, Cubans in both cities have creatively and carefully reimagined their environments to reflect and maintain particular elements of Cuba and their connection to the island, be it ongoing or in the distant past. Howard Murphy (1995:187) has characterized such activity as "the sedimenting of history and sentiment in the landscape." Cuban identity in these cases has been displaced from the island into the diaspora, resulting in an effort to hold on to particular parts of one's identity in specific ways as they integrate into a new place. That "holding on" results in a sense of connection to the island—a sense of hope—when living there is no longer possible.

It stands to reason that Cuban exiles' nostalgia and memory in the United States are also deeply influenced by their contemporary surroundings; this represents a significant disparity between Miami and Moultrie. Miami is a bustling international city, which is generally proud to display its Cuban influences, and public cultural displays (e.g., festivals, place names, and Spanish language media) are common. Miami's Cuban community (and even people outside of that community) counts shrines, cigar stores, parks, restaurants, and monuments among the places contributing to its transnational landscape that incorporate elements of Cuban and U.S. culture. Moultrie, on the other hand, scarcely notices its Cuban population, and Cuban exiles are not especially eager to make themselves known, owing in part to the community's negative responses to other immigrants in the region. Instead, they quietly carry on, away from the prying eyes of vaguely suspicious locals.

Miami, along with other historically Cuban American cities in the United States, has a longer and more complex history of interaction with the Cuban exile community than does Moultrie, which affects the level of political and socioeconomic security experienced by immigrants at each site. The presence of a social network upon arrival at a migration destination is a key indicator of migrant "success" (Palloni et al. 2001). While Moultrie Cubans take advantage of social networking, especially through the Jehovah's Witness community, this network and its resources remains limited compared with those in Miami.

For people in exile or immigrant communities such as these, nostalgia plays a key role in the ways memory is materialized. Moultrie's "landscape of the present" (Stewart 1988:227) is not particularly conducive to public cultural displays from its recent immigrant population. Cubans there have turned primarily to their homes and places of worship as sites at which to re-member familiar or comforting elements of Cuban culture— and they live in and are influenced by that landscape. It is within these more private places that memories are rehashed (in a language that suits them), familiar foods are prepared and shared, gardens are planted and savored, and the freedom to express it all without judgment from the mainstream is most felt. Nostalgia plays no less a role in Moultrie than in Miami; it simply takes a quieter, more private form influenced by the setting and the more recent arrival of this group of exiles.

Consideration of nostalgia as a heightened form of resistance to uniformity and obliteration of cultural identity—even nostalgia for things one did not experience firsthand—can play an important part in the way that displaced or marginalized peoples cope and persist. Moreover, it has a significant bearing on the complementary diasporic movement of viable complexes centered on food and plants from the homeland to destinations around the world. Scholars and activists need to take this aspect of globalization into account as they look at the problem of loss and conservation in more contemporary contexts (see chapter 12, this volume). The transplanting of familiar elements and themes into a new landscape— ranging from cuisine to gardens to architecture—often plays a role in the experience of nostalgia, but it also plays an unexpected role in opening new spaces for diversity to flourish. Thus, reterritorialization, as a partner and product of rememberance, enhances and extends cultural and biological diversity in multiple ways, in both private and public spheres.

Notes

1. Owing to the sheer quantity of these popular and academic publications, I shall keep my discussion of Miami here somewhat brief.

2. Names throughout this chapter have been changed in accordance with University of Georgia institutional review board guidelines and interviewee preferences.

3. Notably, interviewees in Miami nearly uniformly preferred to speak with me in English, while those in Moultrie spoke Spanish.

References

Abercrombie, Thomas A. 1998. *Pathways of Memory and Power: Ethnography and History among an Andean People*. Madison: University of Wisconsin Press.

Airriess, Christopher, and David Clawson. 1994. Vietnamese market gardens in New Orleans. *Geographical Review* 84(1):16–31.

Alston, Jon P., and B. E. Aguirre. 1970. Congregational size and the decline of sectarian commitment: The case of the Jehovah's Witnesses in South and North America. *Sociological Analysis* 40(1):63–70.

Appadurai, Arjun. 1990. Disjuncture and difference in the global economy. *Public Culture* 2(2):1–24.

———. 1996. *Modernity at Large: Cultural Dimensions of Globalization*. Minneapolis: University of Minnesota Press.

Armstrong, H. 1999. Migrants' domestic gardens: A people-plant expression of the experience of migration. In *Towards a New Millennium in People-Plant Relationships*, edited by M. D. Burchett, J. Tarran, and R. A. Wood, 28–35. Sydney: University of Technology.

Basso, Keith. 1996. *Wisdom Sits in Places*. Albuquerque: University of New Mexico Press.

Behar, Ruth, and Lucía M. Suárez, eds. 2008. *The Portable Island: Cubans At Home in the World*. New York: Palgrave.

Berliner, David C. 2005. The abuses of memory: Reflections on the memory boom in anthropology. *Anthropological Quarterly* 78(1):197–211.

Casey, Edward S. 1997. How to get from space to place in a fairly short stretch of time: Phenomenological proglegomena. In *Senses of Place*, edited by Stephen Feld and Keith Basso, 13–52. Santa Fe, NM: School of American Research.

———. 1998. *The Fate of Place*. Berkeley: University of California Press.

Choo, S. 2004. Eating satay babi: Sensory perception of transnational movement. *Journal of Intercultural Studies* 25(3):203–212.

Christensen, P. 2001. Mac and gravy. In *Pilaf, Pozole, and Pad Thai*, edited by S. Innes, 17–39. Amherst: University of Massachusetts Press.

Christie, Maria Elisa. 2004. Kitchenspace, fiestas, and cultural reproduction in Mexican house-lot gardens. *Geographical Review* 94(3):368–390.

Clifford, J. 1994. Diasporas. *Cultural Anthropology* 9(3):302–338.

Cobas, José A., and Jorge Duany. 1997. *Cubans in Puerto Rico: Ethnic Economy and Cultural Identity*. Gainesville: University Press of Florida.

Colquitt County Community Assessment. 2007. Camilla, GA: Southwest Georgia Regional Development Center.

Connerton, Paul. 1989. *How Societies Remember*. Cambridge: Cambridge University Press.

Croucher, Sheila L. 1997. *Imagining Miami: Ethnic Politics in a Postmodern World*. University Press of Virginia.

Dawdy, Shannon Lee. 2002. "La comida mambisa": Food, farming, and Cuban identity, 1839–1999. *Nieuwe West Indische Gids/New West Indian Guide* 76(1–2):47–80.

Duany, Jorge. 2005. La migración cubana: Tendencias actuales y proyecciones. *Encuentro de la Cultura Cubana* 36:164–179.

Dusselier, Jane. 2002. Does food make place? Food protests in Japanese American concentration camps. *Food and Foodways* 10(3):137–165.

Escobar, Arturo. 2001. Culture sits in places: Reflections on globalism and subaltern strategies of localization. *Political Geography* 20:139–174.

Feld, Stephen, and Keith Basso. 1996. *Senses of Place*. Santa Fe, NM: School of American Research.

Garcia, Maria Cristina. 1996. *Havana, USA*. Berkeley: University of California Press.

Glick-Schiller, Nina, Linda G. Basch, and Cristina Szanton Blanc. 1992. *Towards a Transnational Perspective on Migration: Race, Class, Ethnicity, and Nationalism Reconsidered*. New York: New York Academy of Sciences.

Gordillo, G. R. 2004. *Landscape of Devils: Tensions of Place and Memory in the Argentinean Chaco*. Durham, NC: Duke University Press.

Graham, Sonia, and John Connell. 2006. Nurturing relationships: The gardens of Greek and Vietnamese migrants in Marrickville, Sydney. *Journal of the Geographical Society of New South Wales* 37(3):375–393.

Imbruce, Valerie. 2007. Bringing Southeast Asia to the Southeast United States: New forms of alternative agriculture in Homestead, Florida. *Agriculture and Human Values* 24(1):41–59.

Levitt, Peggy, and Nina Glick-Schiller. 2004. Conceptualizing simultaneity: A transnational social field perspective on society. *International Migration Review* 38(145):595–629.

Logan, J. R., R. D. Alba, and T. L. McNulty. 1994. Ethnic economies in metropolitan regions—Miami and beyond. *Social Forces* 72(3):691–724.

Mahler, Sarah J. 1995. *American Dreaming: Immigrant Life on the Margins*. Princeton, N.J.: Princeton University Press.

Mankekar, P. 2002. "India shopping": Indian grocery stores and transnational configuration of belonging. *Ethnos* 67(1):75–98.

Massey, Doreen. 1994. *Space, Place, and Gender*. Minneapolis: University of Minnesota Press.

McDowell, L. 2004. Cultural memory, gender, and age: Young Latvian women's narrative memories of war-time Europe. *Journal of Historical Geography* 30:701–728.

Murphy, Howard. 1995. Landscape and the reconstruction of the ancestral past. In *The Anthropology of Landscape: Perspectives on Place and Space*, edited by Eric Hirsch and Michael O'Hanlon, 184–209. Oxford: Clarenden Press.

Nagel, C. R. 2002. Geopolitics by another name: Immigration and the politics of assimilation. *Political Geography* 21(8):971–987.

Nazarea, Virginia D. 2005. *Heirloom Seeds and Their Keepers*. Tucson: University of Arizona Press.

———. 2006. Local knowledge and memory in biodiversity conservation. *Annual Review of Anthropology* 35:317–335.

Ong, Aihwa. 1999. *Flexible Citizenship: The Cultural Logics of Transnationality*. Durham, NC: Duke University Press.

Palloni, Alberto, Douglas S. Massey, Miguel Ceballos, Kristin Espinosa, and Michael Spittel. 2001. Social capital and international migration: A test using information on family networks. *American Journal of Sociology* 106(5):1262–1298.

Pérez, Luis A. 1978. Cubans in Tampa: From exiles to immigrants, 1892–1901. *Florida Historical Quarterly* 57(2):129–140.

Portes, Alejandro. 1987. The social origins of the Cuban enclave economy of Miami. *Sociological Perspectives* 30(4):340–372.

———. 1997. Immigration theory for a new century: Some problems and opportunities. *International Migration Review* 31(4):799–825.

Rappaport, J. 1998. *The Politics of Memory: Native Historical Interpretation in the Colombian Andes*. Durham, NC: Duke University Press.

Ray, K. 2004. *The Migrant's Table: Meals and Memories in Bengali-American Households*. Philadelphia: Temple University Press.

Rigney, A. 2005. Plentitude, scarcity and circulation of cultural memory. *Journal of European Studies* 35(1):11–28.

Sanders, J. M., and V. Nee. 1987. Limits of ethnic solidarity in the enclave economy. *American Sociological Review* 52(6):745–767.

Seremetakis, C. Nadia. 1996. *The Senses Still: Perception and Memory as Material Culture in Modernity*. Chicago: University of Chicago Press.

Skop, Emily H. 2001. Race and place in the adaptation of Mariel exiles. *International Migration Review* 35(2):449–471.

Smith, Michael P., and Luis Guarnizo. 1998. *Transnationalism from Below*. New Brunswick, NJ: Transaction.

Stewart, Kathleen. 1988. Nostalgia—a polemic. *Cultural Anthropology* 3(3): 227–241.

Stoller, Paul. 2008. *The Power of the Between: An Anthropological Odyssey*. Chicago: University of Chicago Press.

Suárez, Lucía. 2006. *The Tears of Hispaniola: Haitian and Dominican Diaspora Memory*. Gainesville: University Press of Florida.

Sutton, David. E. 1998. *Memories Cast in Stone: The Relevance of the Past in Everyday Life*. New York: Berg.

————. 2001. *Remembrance of Repasts: An Anthropology of Food and Memory.* Oxford: Berg.

Tweed, Thomas A. 1997. *Our Lady of the Exile: Diasporic Religion at a Cuban Catholic Shrine in Miami.* New York: University of Oxford Press.

U.S. Census Bureau. 2000. U.S. Census. Washington, DC: U.S. Department of Commerce.

Vertovec, Steven. 1997. *Migration and Social Cohesion.* Cheltenham, UK: Edward Elgar.

Wimmer, Andreas, and Nina Glick-Schiller. 2003. Methodological nationalism, the social sciences, and the study of migration: An essay in historical epistemology. *International Migration Review* 37(3):756–610.

CHAPTER TWELVE

When Seeds Are Scarce

Globalization and the Response of Three Cultures

ROBERT E. RHOADES

With seeds understandably being crucial to survival and identity, farming and gardening societies always maintain a strong cultural and agronomic link between their planting material and the rest of their food system. Survival depends on the ability of the farm household to carefully select the appropriate material from which the next harvest will come. Thus, seed procurement, selection, storage, and preparation often embody detailed local knowledge of plants and environment. The preplanting process also carries social and ritual significance for most subsistence producers and plays a central role in origin myths and legends (see chapter 4, this volume). Over the centuries, smallholder farmers and gardeners have learned to diversify their agriculture systems and utilize informal seed networks to make sure that seeds are viable and available when needed.

Even in modern agrarian systems serving growing urban populations, the importance of making sure clean, healthy seed is available to producers in a timely way is no less important than in the past. Despite historic demographic shifts away from family farms and a concomitant downgrading of agriculture as a way of life, food and fiber production remains central to human civilization as we know it. In recent decades, seed production has become a major global agribusiness that is increasingly controlled by a few multinational companies. Breeder-improved materials are delivered to farm operations through centralized formal seed systems managed by governments or corporations instead of local social networks. Scientific research, especially in biotechnology, aims to produce planting materials that will further increase global production while guaranteeing greater profits to seed and biotechnology companies.

Plant genetic resources in the form of traditional landraces and ancestral wild species of food crops have become the flashpoint in a heated international struggle over the legal and political control of seeds (see chapter 10, this volume). These seeds comprise the raw input for conventional breeding as well as bioengineering of commercial hybrids and lines. Biocapitalistic corporations of the North attempt to control their investments in seed technology through intellectual property instruments and genetic modifications (e.g., terminator gene), while post-Convention on Biological Diversity, gene-rich countries of the South and their indigenous communities have officially shut the door to bioprospecting by outsiders. Thus, in the shift from smallholder agriculture to corporate farming, seeds have not become less important. The main difference, and the center of the current debate, is who controls access to the seeds crucial for improved crop production.

In the rush to protect what different parties in the North-South "seed wars" feel is rightfully theirs, it is often overlooked that more marginal, small-scale local curators of landrace biodiversity outside of the formal sector increasingly face situations of seed scarcity. The causes of this local scarcity are numerous, including environmental disruptions such as drought and flooding, population dislocations, wars, pests and diseases, and economics. One underlying force generating genetic and plant knowledge erosion among small-scale producers is globalization, the accelerating trend toward an economically, culturally, and politically homogenized world (Appadurai 2001). The number of smallholder farms throughout the world continues to decline as rural people move to cities seeking wage employment. Food becomes a commodity to be purchased and no longer an integral part of the household production-consumption cycle. The Green Revolution of the 1960s and 1970s penetrated most local agricultural systems with the consequence that traditional landraces were discarded in favor of introduced hybrid varieties from national breeding programs or multinational seed companies. Farmers in these contexts, while initially captivated by introduced "improved" and "miracle" seeds, soon experienced the genetic vulnerabilities and economic risks of specialized breeding lines (see Introduction, this volume). Unfortunately, if there was a desire to return to their traditional varieties, farmers often discovered that they were hard to find or had completely disappeared.

Another manifestation of globalization is international migration wherein people either voluntarily or involuntarily leave their homelands for foreign countries. Well over 120 million people have in recent decades become immigrants or refugees, taking with them not only their energy and hopes to their new home but also their food habits. In their adopted countries, reflecting on the old adage that the last thing to change in an immigrant's home is the cupboard, immigrants seek to grow or obtain plants essential to their native cuisine. Aside from fulfilling food memories, gardening also performs another function, namely, to help immigrants set down roots in the new place. The seeds of the immigrants' culture, however, are typically unavailable or, at best, scarce in the new land.

In this chapter, I examine the dynamics of recovery, repatriation, and *in situ* conservation of culturally relevant plants among three distinct globalizing populations. I discuss how globalization has affected local agrobiodiversity, the ways in which gardeners and farmers have responded to maintain their genetic resources, and the impacts of conservation projects. The three cultures included here share little more than the influence of globalization and a common desire to revive, maintain, and use endangered or scarce traditional plants. Southern U.S. heirloom seed keepers represent a marginal population within the global core; Vietnamese immigrant and refugee gardeners in the United States are attempting to recreate a plant complex from their homeland; and indigenous farmers of Cotacachi, Ecuador, are struggling under the pressure of globalization to maintain ancestral crops in a center of crop origin and diversity.

Seed Keepers of the American South: Marginality in the Global Core

Many Americans believe the most recalcitrant and culturally conservative region in the United States is the American South. Broadly defined, this region is made up of eleven states of the Old Confederacy (Virginia, North and South Carolina, Georgia, Alabama, Tennessee, Mississippi, Arkansas, Texas, Florida, and Louisiana), along with adjacent regions of Kentucky, Maryland, Missouri, and Oklahoma (Pillsbury 2006). Blessed with abundant sunshine and an average of fifty inches of annual rainfall, the American South has supported some of the most productive and diverse agriculture in the United States. Although the popular image is an agrarian landscape defined by rural tobacco farms and slave-based cotton plantations, the region in reality has always been made up of a mix of large-scale operations and small, diversified family farms and homegarden patches.

In many ways and despite its rapid transformation, the South remains America's "margins," its less developed region. Southerners are still undeservedly stereotyped and ridiculed as backward by outsiders (Peacock et al. 2005). Although located in the very nation that defines globalization, the socioeconomic forces transforming the South over the past fifty years are strikingly similar to those experienced by developing countries. Small-scale farming began to disappear after World War II as black and white Southerners moved away from family farms and sharecropping relationships to find employment in cities and factories. The rural population declined by more than half after 1960, a demographic shift that began a process of emptying the countryside of its long-term inhabitants. Even if families stayed in farming, they increasingly gave up production for household consumption in favor of selling all farm products and then purchasing their own food in stores. From the 1960s onward, government agencies, land grant colleges, and agribusiness promoted adoption of modernized, scientific farming equipment and methods. Those who remained on the farm and had access to credit through the Farm Bureau and other farm support programs enlarged their landholdings and mechanized production. Farming thus "became a business, leaving behind farming as a way of life" (Hurt 1998:26).

Another force creating the "New South" was the influx of people from other U.S. regions and foreign countries. Although the out-migrations to the industrial North dominated the Southern story from the 1930s until the 1970s, a counterflow of people poured into the region attracted by jobs and the "sunbelt" climate. Atlanta—the hub of the South—epitomized this growth as it became known as a global telecommunications center (CNN), home of the world's busiest airport (Hartsfield-Jackson International Airport), and headquarters to the ultimate symbol of globalization (Coca-Cola). As in the rest of the United States, the processes of Wal-Martization, McDonaldization, and suburbanization took root throughout the region. On the agricultural front, farming transformed from family-run operations to agribusiness, the latter often based on a factory farming model. This model involves capital-intensive animal raising in confined animal feeding operations. With labor and environmental protection laws and weak unions favorable to employers, the South became a mecca for multinational poultry and swine operations staffed largely by immigrant labor from Mexico and Central America.

Some Southerners today speak about a "period of shame" in which they wanted to shed their regional accents and mannerisms so they would be accepted by mainstream Americans. Any connection with farming or rural life was seen as part of the less esteemed Southern identity. At the same time, land grant universities and state extension services pushed modernization of farming and the shift away from local plant varieties and animal breeds to "improved" ones. State governments began to regulate seeds in ways which did not encourage informal and independent seed distribution. For example, farmers in Georgia had relied since 1917 on the *Farmers Market Bulletin*, a rich information source on farm items for sale or exchange published by the Georgia Department of Agriculture. Seeds of garden heirlooms were exchanged and sold through the bulletin, which reached farm families by post each Thursday. Back in 1936, there were no published restrictions on sale or exchange of seeds. Until the 1990s, the bulletin was a joy to read for those interested in diversity of both plant varieties and animal breeds. By the mid-1990s, however, to advertise seed for sale one had to submit a self-reported germination rate less than 9 months old on a form provided by the government. By early 2000, if a gardener wanted to place an ad to sell heirloom seeds, a certified laboratory report for purity, noxious weeds, and germination rates must be provided. Subscribers are also admonished that seed lots offered had to be uniform, of a certain size, and not in violation of plant protection laws. As can be expected, today fewer old-timey seeds are advertised or swapped in the *Farmers Market Bulletin* (www.agr.geor gia.gov).

Another example of the state's role in fostering genetic erosion concerns Southern apple varieties. Southern families had grown apples since they arrived from Europe, and by the nineteenth century the fruit had become a major source of income for orchard owners. In 1930, for example, Georgia had one million apple trees in commercial production, not counting the ubiquitous apple trees found around the homeplaces of the South. Today, only a handful of apple orchards operate and with few distinct varieties. Of 1,400 or so apple varieties once grown in the South, only about two hundred exist today, and over 90 percent of these are grown by a few dedicated lay "apple keepers." According to Lee Calhoun (1995), an expert on Southern apples, one of the finest collections of old-timey apples was found at the University of Georgia's Mountain Research Station in Blairsville, Georgia. In the 1970s, Georgia began to import most of its apples, mainly Red Delicious, from the commercial orchards in Washington State. Deeming the old apple varieties like Yates, Etowah, and Rabun useless, the university sent out workers with chainsaws to dispense with their large collection of heritage apple trees. A small band of independent Southern apple hunters and growers heard about the chain saw massacre and rushed to the farm to save what they could. Ironically, today there is a strong desire among Southern consumers to find the old apple varieties. Each fall, thousands of urbanites from Atlanta and other cities drive to the mountains in search of apple farms and festivals as a way to experience a variety of apples and a prior way of life.

Despite strong formal and informal forces working against heirloom varieties, they have continued to persist and even flourish in certain geographical pockets and social spaces in the South. Because of economic necessity, geographical isolation, better taste or cooking quality, or simply delight with diversity, Southerners have maintained, shared with family and friends, and passed along their collections. In her book *Heirloom Seeds and Their Keepers* (2005), Virginia D. Nazarea describes the essence of Southern heirlooms:

> Signifying not only age but, more importantly, lineage and legacy, heirloom plants are highly prized by many Southerners, whether one is talking about garden peas, apple trees, or antique roses. Two widely agreed upon criteria are used to distinguish heirloom plants: first, they have been around for fifty years or longer; and second, they are open pollinated and can be propagated by saving and sowing seeds from a previous harvest. A third quality, which is not always required, is that they have been passed down, or passed along, a kinship-based or other informal or folk network and thus have existed for an extended period of time outside formal seed systems and breeding circles. (80)

Throughout the South, more than any other American region, one finds a landscape still defined by homegardens and orchards near the house. These are typically found in rural areas isolated from interstate highways and strip malls, but sometimes they flourish in the heart of booming cities. A good example of the latter is Ernest Keheley of Marietta, Georgia. Marietta is a large upscale suburb of Greater Atlanta, whose residents tend to be rich and conservative. Newt Gingrich, former majority leader of the U.S. Congress, purposely moved to Marietta to build his political base because it has one of the highest per capita incomes and one of the greatest concentrations of conservative Republican values in the United States. But in the midst of the McMansions there, Mr. Keheley lives in his family's homeplace, a 100-year-old Victorian house with a wraparound porch, where he sits in a rocking chair and waves at neighbors passing by in their SUVs and high-end sedans. Probably wealthy himself because he sold much of the family farm to developers, Ernest was still an avid Southern gardener. Up until the early 1990s, when he suffered a stroke, he planted a large garden annually and maintained a personal collection of dozens of rare heirlooms. His approach was hardly what modern agriculturists would find rational. For example, in an old coffee jar he keeps a seed mixture of Hickory King Corn, Hastings Field Corn, and three or four varieties of beans. He says the mixture is more efficient than just planting one type; the corn stalk is the pole for the different beans, which yield abundantly at staggered times. Other heirlooms in his collection included White Crowder peas, Big Boy pea, Martin gourd, "Whipperwill" Crowder pea, Hercules pea, Red Ripper pea, Pink Eye Pea Colossus, Pink-Eye Purple Hull, Blue Goose bean, Knuckle Hull Cooper pea, Plumgranny, and many more. For each of these varieties, Mr. Keheley has a personal or family story.

The South prides itself on the distinctiveness of its regional culture, defined not only by its literature, art, and music but also by its cuisine. Life for many Southerners still revolves around family with cooking and food gatherings as central social activities. Typically called "pass-alongs," old-timey varieties are maintained by many families as ingredients in the distinct Southern cuisine (Bender and Rushing 1993; Nazarea 2005). Heirlooms are valued for their aroma, flavor, and specific roles in different dishes. For example, the collard patch one sees in the fall and winter next to many a Southern home is a cultural icon. The geographical distribution of collards more or less delineates the boundaries of the South. The collard plant (*Brassica oleracea*) is adapted to the summer heat and, with its high vitamin A and C, calcium, and iron content, is valued for its highly nutritious leaves. Southern growers are passionate about growing the right variety (e.g., Cabbage collards, Georgia collards). Collards are served as a side dish and often with sweet potatoes, black-eyed peas, and corn bread. Cooked collards produce a juice called "pot likker" that is used to soak corn bread (Davis and Morgan 2005).

When Virginia D. Nazarea and I came to Georgia in the early 1990s, we realized that this region offered a unique opportunity to study and help preserve traditional landraces along the lines of the applied research we had been doing internationally with rice, potatoes, and sweet potatoes. The marginality and cultural resistance in the South turned out to be, in fact, a plus for preservation of old-timey plants and their uses. We knew from our work in Asia and the Andes that marginal farm families maintain greater landrace agrobiodiversity than those linked closely to roads, markets, and urban areas (Rhoades and Nazarea 1998; Gepts 2006). We also knew that as a general rule the elderly in a community were the most knowledgeable about these plants and that most communities had at least one passionate seedsaver whose self-imposed lifetime passion was to multiply and save the diversity of local varieties (Nazarea 2005). We found these same patterns of recalcitrance and idiosyncrasy in the American South.

Perhaps because of the viability of informal cultivation and exchange of heirloom seeds, no formal seedsaver organizations had developed in the South similar to those of other American regions. This is most likely because special organizations were not needed to help Southerners do what they are already doing quite well. On the other hand, we knew that even with the countervailing forces of Southern identity and culture working in favor of *in situ* landrace conservation, rampant agricultural commercialization made the situation tenuous at best. In early 1990, when we first became interested in Southern heirlooms, Southerners seeking Georgia Rattlesnake watermelon (an old-timey Southern variety) found themselves writing to northern addresses like Decorah, Iowa, where Seed Savers Exchange would send them a twenty-five-seed packet of a Southern seed for a fee. Today, old-timers recount moving memories of a past with much more diversity and they lament the loss of certain varieties. With this as motivation, we founded the Southern Seed Legacy (SSL), a regional grassroots seedsaving and exchange network similar to Native Seeds SEARCH in the American Southwest, in 1994. Since then, the SSL has collected almost eight hundred named varieties, along with memories and local knowledge of their characteristics and uses, and have circulated these among experienced and novice gardeners alike through seed swaps and an initiative called PASS (for Pass-Along Southern Seed). One retired gentleman contacted the SSL looking for Hickory King corn. He explained that this corn variety was planted by his Mama during the Great Depression and saved the family from starvation. He wanted to find Hickory King seeds and grow them out "just for memory's sake." Younger Southern gardeners are constantly searching for old named varieties they once tasted and heard about from friends and family.

Our SSL research uncovered a quiet but highly viable community of heirloom or "old-timey" gardeners who still passionately cultivate distinctly Southern plants. For the most part, they are not connected with any formal seedsaving or organic farming organization. Most are direct descendants of farmers and gardeners rooted in the South who have saved the seeds for generations as a matter of course. While many Southerners save a variety of heirlooms in their gardens, some specialize in one species. In Oklahoma, we visited Carl Barnes, the "Corn Man," who had one of the largest open-pollinated maize collections in the world stored in his double-wide trailer house (see figure 12.1). In Alabama, we learned through SSL participant Susannah Chapman about Joyce Neighbors, an eighty-five-year-old "apple keeper" who collects, cultivates, and sells tree saplings of Southern heirloom apples (see chapter 2, this volume). Then, there is Roger Winn, the "Tomato Man" from Little Mountain, South Carolina, who grows dozens of distinct tomato varieties and sells seedlings in our annual seedswap only to donate the income generated to the support of SSL.

While most seedsavers are private, low-key individuals, a few see a larger cause in saving Southern seeds. On the Georgia coast in Liberty County, Eleanor Tison, another SSL participant, interviewed John Stevens, an African American man who has revived the nearly forgotten

FIGURE 12.1. Carl Barnes, a seedsaver from Oklahoma, maintains a large collection of open-pollinated landrace corn. Photo by Robert E. Rhoades.

Seminole pea (Péralte 2003). On New Year's Day in the South, one sits down to a plate of Hoppin' John—black-eyed peas and collard greens. This combination is supposed to bring good luck: the greens signify dollars while the peas are copper-like pennies. However, long before the black-eyed pea got into this traditional dish, Hoppin' John was made from collard greens and the Seminole pea. The Seminole had nearly disappeared after the old-timers in his area had died off and young people had migrated away. But John Stevens had acquired a few seeds, planted them, and luckily reaped a small harvest. The Seminole pea is only grown in Liberty County, and John hopes to duplicate the place-of-origin success story of Vidalia onions grown in Tombes County, Georgia, which earns $90 million annually from the sale of the famous onion that is reputed to develop its unique sweetness only on Georgia soil. In fact, the canning company Sunburry is canning Hoppin' John made with Seminole peas, and the proceeds go to Seabrook Villages, a living-history African American museum on the Georgia coast and part of the Black Heritage trail that runs from Florida to the North (Péralte 2003).

The Southern milieu, including its cuisine, dialects, landscapes, music, and social connections, is a distinct and cherished place-based culture. The traditional cuisine of corn bread, collard greens, barbeque, ham hock and beans, and dozens of other uniquely Southern dishes is alive and well. Southern gastronomy places a high premium on having the right landrace ingredient. Compared with the rest of the United States, strong kinship, religious, and community values still flourish, with ceremonies and rituals tied to social gatherings such as taffy pulls and sorghum syrup making. In sum, saving Southern landraces from extinction is driven not by economics but by culture, place, and memory. The homogenizing forces of globalization are countered by a desire to link with the past, the landscape, and tastes born of family and friends.

Vietnamese Gardeners in the American South: Transplanting the Homeland

If the South had been relatively isolated from the rest of the United States until the last half of the twentieth century, it was even more isolated from the rest of the world. Rapid globalization of the South, however, corresponded not only with the dramatic rise of Fortune 500 Southern companies such as Bank of America and Wal-Mart but also with the arrival of unprecedented numbers of immigrants. Although Hispanic migrants predominate—migrants from Mexico and Central and South America working in construction, meat processing, landscaping, and carpet factories—millions also arrived from Asia, Africa, and Eastern Europe. As immigrants come to seek a new life, they bring with them a cultural scaffolding that has distinct and persistent food habits at its base. Food of the home country is an important link with the past and a marker of ethnic identity and is the part of culture most resistant to change. While immigrants have few options in coping with the strangeness of a new language, confusing laws, discrimination, and racism, one area of their lives they can control is what they cook in their own kitchens and eat in the company of compatriots. Cultivation of familiar plants, cooking of favorite dishes, and partaking of meals with close family and friends help cushion the many unsettling encounters of immigration (Kalcik 1984).

The Vietnamese are among the most recent immigrant groups to the United States who today number more than 1.2 million. They came in waves: first, in 1975 fleeing the communist Viet Cong's takeover of South Vietnam; subsequently, in the 1980s as political refugees; and from the 1990s until the present, as voluntary immigrants under the present immigration policy. They live in dispersed clusters throughout the United States, but the largest populations are in California, Texas, Louisiana, Florida, and Georgia. Like other Asian immigrant groups, the Vietnamese are portrayed as a "model minority," yet they have suffered significant social and psychological strains, alienation, intergenerational conflicts, low incomes, and poor housing (Do 1999:14, 109–110). Members of the first wave of Vietnamese immigrants (1975–77) are gaining economic parity with other Americans, but the economic status and social mobility of subsequent groups remain below the national average. Vietnamese have tended to settle near previously established Vietnamese communities, leading to ghettoization, or in neighborhoods characterized by diverse immigrant groups. One example of the latter pattern is Buford Highway near Atlanta, Georgia, where Latinos and Asians have created a new immigrant economy and landscape. Buford Highway supports migrant-targeted businesses such as supermarkets, travel agencies, real estate companies, banks, and music and video rental shops with foreign language films, and ethnic restaurants. Vietnamese restaurants serving *pho* and other Vietnamese foods successfully cater to a broad range of clientele, including other immigrant groups and Afro- and Euro-Americans. Most customers, however, are Vietnamese who find comfort in settings where they speak their language, hear their music, and enjoy the tastes and smells of the homeland.

The centrality and adaptive functions of food for immigrant groups are well documented in the social science literature (Kalcik 1984; Shortridge and Shortridge 1998). Food symbolically marks social boundaries and helps immigrants reaffirm their own culture within an alien one (Airriess and Clawson 1991; Kalcik 1984). Since many Vietnamese in the United States are dispersed and socially marginal, food becomes a focal point in gatherings to celebrate holidays and religious ceremonies and to honor the ancestors. Traditional Vietnamese cook favorite foods of the deceased on the anniversary of their deaths and leave dishes on an altar to the ancestors (Kalcik 1984). Hospitality in Vietnamese culture is defined in terms of food. A Vietnamese will not ask a visitor, "How are you today?" but rather, "Have you eaten today?" (Owens 2003:39).

Vietnamese cuisine requires particular spices, herbs, fruits, and fresh vegetables, which are often prepared with meat. Faced with a new environment devoid of these key plant ingredients, however, Vietnamese families in America seek mechanisms and connections to acquire seeds or planting materials necessary to recreate their traditional cuisines. While ethnic supermarkets are increasingly addressing the cultural food needs of immigrant groups, many of the specialized vegetables or herbs required for Vietnamese dishes cannot be obtained in a general Asian supermarket that is more than likely Chinese or Korean. The primary mechanism, therefore, of acquiring appropriate cooking ingredients in the United States is gardening, which in turn requires space, knowledge, and the essential planting material. We also found that even if the plants are available in Asian stores, the homegarden provides readily available, fresh, and often organic produce to the family and their friends.

The widespread presence of new immigrant gardens and the exceptional diversity of plants they display caught the attention of Henry Shands, director of the National Center for Genetic Resources Preservation located in Fort Collins, Colorado. The center is the U.S. government agency in charge of assessing, acquiring, and maintaining genetic resources of value to the United States. Realizing that a great deal of new domesticated plant biodiversity was being introduced by gardening immigrants about which little was known, Henry Shands inquired if Virginia D. Nazarea and I would be willing to conduct a study of the Vietnamese plant introductions with the possibility of expanding the research to other recent immigrant groups. Through funding from the U.S. Department of Agriculture (USDA) Plant Exploration Unit, our project, titled Introduced Germplasm from Vietnam: Documentation, Acquisition, and Preservation, aimed to document recently introduced Vietnamese germplasm to the United States within a cultural context (Rhoades and Nazarea 2003). To implement this research, we recruited Vietnamese American college students in Georgia to conduct interviews and collect seeds and other planting materials of Vietnamese vegetables, fruits, herbs, and ornamentals. Using the memory banking approach (Nazarea 1998), we documented plant uses, acquisition of planting materials, gardening, health and healing practices, and beliefs associated with the plants for use in cooking and ceremonies.

Vietnamese gardens in the United States have diverse functions, providing an important source of food, income for some individuals, and healthy physical activity (Owens 2003; Do 2003). If immigrants have a good harvest, they can share the surplus, which in turn builds social networks. Lynn Do (2003), one of our student participants, found that gardening gave her informants joy, simply from the pleasure of growing their own Vietnamese vegetables and herbs. Airriess and Clawson (1991) conducted a study of elderly Vietnamese market gardeners in New Orleans and concluded that gardening was a form of horticultural therapy. For these elderly immigrants, gardening was a purposeful strategy to help them cope with the challenges of a strange society, as well as their changing roles within their own culture, and provided excellent nutrition through familiar foods. Gardening increases the elderly's sense of self-worth and gives them some degree of independence from the younger generation in America (Rhoades and Nazarea 2003).

Vietnamese gardening in America is also about recreating place and memory. Gardens are designed to be physical "refuges" within which Vietnamese deal with the stress of being uprooted and living in a foreign land. Immigrants want to recapture a memory landscape of Vietnam. As Lynn Do (2003:21) in her study of Fort Myers, Florida, noted, gardens help "bring a feeling of Vietnam to their lives." One of her informants declared: "I garden because I want my house and yard to look like Vietnam." Richard Owens (2003) reported that gardeners in Lincoln, Nebraska, select homes based on their gardening potential. In both places, an important reason for preferring home ownership over renting is the freedom to plant their own gardens. Moreover, the number of species and garden stability increase with years of ownership.

Recent Asian immigrant homes in the American South are notable for the unique pattern of landscaping. Some immigrant families try to keep a low profile by maintaining a traditional front lawn with decorative shrubs and flowers, but more often than not the backyard tropical garden spills into plain view. The most distinctive feature of the Vietnamese garden is the vertical layered production system (see figure 12.2). As in the tropics, plants are cultivated in association to take advantage of sun, shade, support, and water. For example, plants like *cai be xanh* (mustard greens, *Brassica juncea*) might be planted under a sugar apple tree (*Annonasquamosa*). Long beans (*daudua*) and bitter gourds (*bau*) might be trellised together along with other climbing plants. At ground level, one typically finds vegetables like *rau ma* (pennywort), *dau-leo* (cucumber), *dau gang* (similar to watermelon), and herbs for flavoring soups and noodle dishes, like pho (Nazarea 2005). Fruit trees such as avocado (*traibo*), persimmon (*hongvuong*), lychee (*traivai*), papaya (*dud*

FIGURE 12.2. Trellised, staked, or hilled up, Vietnamese vegetables reconstruct place in an adopted homeland. Photo by Robert E. Rhoades.

u), jack fruit (*traimit*), and pomelo (*sa-bo-che*) may be spaced around the yard. Half of all households in Lynn Do's (2003) sample cultivate *bun got* (*Sauropus androgynus*), a bush whose leaves are used to make soup, while all households grow pomelo (*traiboui*), a large, pear-shaped grapefruit. They also grow a wide variety of flowering and medicinal plants. Because of the abundant and layered nature of Vietnamese gardens, there is less concern with weeds compared to the typical American garden. They use old bathtubs, sinks, kiddy pools, five-gallon plastic buckets as containers for water-loving species.

Vietnamese gardens sampled in the United States seem to be as diverse as in Vietnam. Lynn Do (2003) documented ninety-one species in eight separate households, with an average garden consisting of twenty-nine species. A study of rural women in Vietnam documented only sixty-two species of plants (half of which were cultivated, with the rest gathered wild), with an average diversity of five species per garden (Ogle et al. 2001). As gardeners experiment with plants and substitute new plants with more desirable traits, the homegarden becomes a living and evolving site of *trans situ* biodiversity conservation. Vietnamese immigrant gardens are neither *in situ* in the conventional sense (since plants have not evolved in this context or grown for long) nor *ex situ*, as in the sense of a genebank. As used here, the term *trans situ* signifies bringing plant genetic resource materials from a historic *in situ* context (Vietnam) and nurturing them in a completely new national and cultural space (United States) separate from the formal seed sector and while maintaining a continuous genetic infusion from the homeland. It comes down to *transplanting the homeland*, or much-loved parts of it, and living transnationally in very concrete terms. Other Southeast Asian immigrants, like the Thai, have introduced the Vietnamese to other plants, such as *Leucaena leucocephala* (a tree whose pungent leaves and beans are consumed). If they cannot find a familiar plant species from Vietnam, they will substitute local or other introduced species. Although hurricanes and storms have been known to devastate Vietnamese gardens since they like to garden along rivers and canals, the general trend is for garden plant diversity to increase over time.

The Vietnamese gardeners interviewed in our plant exploration project tended not to be farmers or gardeners in Vietnam. Prior to coming to the United States, most immigrants and refugees lived in cities and towns of Vietnam where they purchased their favorite vegetables daily in the local market. Unlike the elderly gardeners from New Orleans studied by Airriess and Clawson (1991), our informants learned gardening after arrival in the United States. Do (2003) found that in Lee Country, Florida, only one of her ten Vietnamese informants had gardening experience in Vietnam. In Lincoln, Nebraska, Owens found that 61 percent of his sample (N = 30) did not garden in Vietnam. Hence, immigrants had to learn not only how to cultivate familiar Vietnamese plants, but they had to do it in unfamiliar surroundings. For the most part knowledge about planting did not come from extension offices, gardening centers, nurseries, or books, or even from home. Instead, they acquired horticultural knowledge from friends and relatives with whom they also exchanged plants and seeds. Some started gardening only after they had been in the United States for a number of years. Early immigrants noted that they did not initially grow any Vietnamese plants because they were unaware of where to get planting material.

Planting material is obtained through five sources: saving one's own seed, exchanging with family and friends, obtaining directly from Vietnam, purchasing seed packets, and extracting from fresh fruits and vegetables bought in local Asian stores. Cuttings are often obtained from Asian stores, as well as from other Vietnamese gardens. For species already growing in their gardens (e.g., sugar apple and longan trees), shoots are started from seeds gardeners collect from their own trees or from those of friends and family. While Asian seed nurseries in California and Taiwan market packets of vegetables and herbs of interest to all Asians, it is likely that more specialized and rarer Vietnamese materials came directly from Vietnam or were exchanged within the United States. Immigrants rarely admit that they bring in seed from Vietnam for fear of government reprisal.

Immigrants had to create their own seed and plant propagation system specifically for Vietnamese germplasm. They keep fruits and vegetables from their harvest and allow these to ripen to maturity, covering with a plastic bag to discourage pests. To save seeds from consumed fruits and vegetables purchased at the market, they air dry them on paper towels, and often distribute the seeds as gifts. Vietnamese women are most often responsible for germinating the seeds and planting the vegetables. Increasingly, tropical nurseries are cropping up in the South to accommodate the new interests. In Florida, an ethnic Chinese woman from Thailand sells grafted lychee, jackfruit, and star apple trees.

The Vietnamese social network is instrumental for exchange of plants and seeds. Sharing seeds in the United States is a way to connect immigrants, sustain their cuisine and identity, and maintain a memory link to the homeland. The seed networks are quite elaborate and crisscross many states. For example, a valued Vietnamese persimmon was first extracted from a fruit bought in a market in California. It was then shipped to a relative in New Orleans, who in turn gave it to a person from Boca Raton, Florida, who then gave it to another Vietnamese immigrant in Fort Myers, Florida (Do 2003). Seeds are exchanged through the mail,

while plant cuttings are exchanged face-to-face among kin at family re-
unions and religious gatherings. Scarcity of seed and planting material
today is increasingly less and less of a problem as the Vietnamese community settles and matures in the United States.

The Vietnamese case illustrates how a people who came involuntarily
to a new land without their traditional plants within the course of a few
decades almost completely recreated the traditional garden crops of the
home country. Acquiring this germplasm was so important they were willing to invest a great deal of energy in social networking focused on acquiring their culturally relevant plants, in working and amending Georgia
red clay, and/or in moving to Florida for a longer more tropical climate
and growing season. We can assume that few seeds were brought over by
the earliest refugees after the fall of Saigon in 1975, but over time the
desire to have their own plants was strong enough to overcome barriers
of legal, political, and biophysical conditions. As among American Southerners, the desire to grow the plants critical for the maintenance of cuisine,
landscape, and memories among Vietnamese immigrants is largely defined
by one's cultural upbringing.

Cotacachi, Ecuador: Conservation
in a Center of Crop Origin

The indigenous Quichua-speaking communities of Cotacachi Canton in
northern Ecuador have a long, rich history of cultivating traditional Andean crops. Like other native groups throughout the high Andes, they
are descendants of ancient farmers who experimented and domesticated
the wild progenitors of potatoes, quinoa, peppers, squashes, and dozens
of other species. Erroneously labeled by modern scientists (e.g., National
Research Council 1989) as "Lost Crops of the Inca," most of these moun-
tainous species remain in cultivation today and play an important role in
the diets and identities of Andean people. These same cultivated plants
supported the Inca civilization, which at its height in the late fifteenth
and early sixteenth centuries spread its influence over more than 2,000
kilometers from Colombia to Chile. Although Cotacachi lies on the
northern margins of the Inca Empire's expansion, Cotacacheños today
consider themselves inheritors of the great Inca civilization.

Today more than 18,000 indigenous people live in forty-one communities scattered around the base of Mama Cotacachi, a 4,939-meter-high
volcano considered sacred by local people (Rhoades 2006a). The climate
varies across a mountain transect ranging from the subtropical inter-
Andean valley floor to alpine conditions above the tree line. Although

less marked than in the past, the zones are still exploited in a complementary fashion by zonally stratified communities that produce crops and
raise animals best adapted to each zone (Moates and Campbell 2006).
Exchange between communities guarantees a diversified diet in all zones.
Households in the high zone higher than 2,700 meters above sea level
graze cattle and sheep, cultivate hardy Andean tubers and grains (potatoes,
mashua, arracacha, melloco, oca, and quinoa), and gather medicinal
plants and wild fruits such as the Andean blackberry (Skarbø 2006:161).
Middle-zone agriculture (2500–2700 meters) is dominated by maize,
beans, squash, wheat, and barley. The warmer low valley zone (2300–2500
meters) supports the highest agrobiodiversity, mainly in household gardens composed of short-cycle, intercropped species, including multiple
landrace varieties of maize, beans, squash, lentils, peas, sweet potatoes,
white carrots, and subtropical fruit like avocado, cherimoya, and passion
fruit (Piniero 2006). Over sixty different food crop species are grown
today in Cotacachi, although about half of these are Old World introductions by the Spanish (Skarbø 2006).

Over the past five centuries, life for indigenous Cotacacheños has
been a legacy of brutal social neglect and disenfranchisement from Ecuadorian society. As late as the 1960s, the Quichuas survived as little
more than modern-day serfs under the hacienda economy of large landholdings of the church or prosperous mestizo families. Under a system
called *huasipungo*, or debt in servitude, indigenous households exchanged
unpaid labor for rights to cultivate small parcels of hacienda land (*huasipungo* means "homegarden" in Quichua) and access to other amenities
such as firewood and water. Despite racism, oppression, and poverty
under the huasipungo system, indigenous Cotacacheños resisted assimilation to mestizo culture over five centuries by maintaining a native cosmology, communal organizations, oral traditions, dress, language, and
crops. Since the hacendados produced mainly commercial crops and animals, local diversity in maize, potatoes, beans, and dozens of minor
Andean crops was kept alive in Quichua gardens and small fields on the
margins of the haciendas. The elderly today recount how, except for salt,
thirty years ago their communities were self-sufficient in food.

In 1963 and again in 1974, Ecuador implemented land reform as part
of its modernization plan. Former hacienda workers were deeded small
landholdings, albeit the least productive land on the hacienda. As in other
parts of the world, land reform corresponded to infrastructure development and technological advances in agricultural productivity. The Green
Revolution in Latin America, as elsewhere, was designed to increase
yields, reduce labor requirements, and hence free "underemployed" rural
labor for employment in other economic sectors. Development agencies
such as the U.N. Food and Agriculture Organization (FAO), the U.S.

Agency for International Development (USAID), the Andean Mission, and Ecuador's Ministry of Agriculture, along with nongovernmental organizations (NGOs), introduced packages of chemical fertilizers, herbicides, pesticides, fungicides, tractors, and input-dependent hybrid seeds. As in Asia, where improved, hybrid rice varieties led to the demise of traditional cultivars, similar losses of Andean varieties occurred in Cotacachi. For instance, where once dozens of potato or maize varieties were grown, by the early 1990s a few modern varieties introduced beginning in the 1970s by the government or seed companies dominated the landscape.

Underlying the loss of Cotacachi's traditional landraces are the same processes of globalization and agrarian change that affect smallholders throughout the world. Steadily, the indigenous people of Cotacachi have been drawn into the wider global economy and communications network. Since 2000, Ecuador has adopted the U.S. dollar as its official currency, and trade liberalization reforms imposed by international donors on the Ecuadorian government are bringing more foreign goods to local markets, including subsidized foods. Development projects, decentralization and democratic reforms, tourism, television, Internet, and improved transportation are now daily influences that put Cotacacheños in contact with the outside world.

Agricultural modernization of the rural sector also led to a need for cash, stimulating young, able-bodied men and women to migrate to Quito and other regions of Ecuador for work in construction, floriculture, textiles, or seasonal agriculture. Despite this out-migration, however, the Quichua maintain enduring familial and land connections to Cotacachi. Between 1960 and 1990 the indigenous population of Cotacachi doubled, leading to further fragmentation of parcels through inheritance and a decline in the ability of rural households to maintain traditional crops and their subsistence needs. Twenty years ago, nearly 100 percent of the produce in the Cotacachi central market was supplied locally by nearby farmers. Today, three-fourths of the produce in the market is imported from other areas of Ecuador and increasingly from other countries such as Chile and Colombia.

More and more women work outside the home and have less time to cultivate their gardens or prepare the "slow" traditional dishes. Preparing the heavier, hearty indigenous meals based on Andean grains and tubers gave way to cooking quicker but less nutritious, light "mestizo" food such as rice and noodles (Camacho 2006). While the elders complain that "light food" (*comida liviana*) purchased in the market today leaves you hungry after meals, the younger generation has developed a preference for store-bought foods (e.g., bread, snacks) because they do not symbolize poverty or Indianness.

Anthony Bebbington (1996, 2001) has drawn attention to "glocalization" in the Andes, a process in which indigenous people have established their own global linkages with the international indigenous movement and sympathetic NGOs to counter concerns about the culturally destructive tendencies of globalization. The central political position of the indigenous movement in Ecuador is that self-determination and preservation of their culture and identity must be treated as centerpieces of development. Given that traditional Andean crops are critical to this cultural identity and play a symbolic role in festivals, rituals, dances, and community reaffirmation, indigenous leaders place a priority on addressing the loss of both plant knowledge and varieties. The indigenous cosmology in terms of time and space is defined by a ritualized agricultural calendar of planting, cultivation, and harvest of crops. These same processes reassert ethnic pride, sense of place, and declaration of territorial rights. To guarantee a bountiful harvest, Cotacachi's shamans will bless fields with an offering of diverse varieties of maize to Mother Earth (*Pachamama*). Inti Raymi, the Inca harvest festival of the Sun beginning at the winter solstice (June 21), requires traditionally prepared meals for community dancers who celebrate and engage in intravillage ritualized conflict as a blood sacrifice to Pachamama or Mother Earth (Wibbelsman 2005). A mildly alcoholic drink called *chica de jora* should be prepared with seven different varieties of maize. During communal gatherings—whether religious, political, or social—each village prepares a feast from the traditional cultivated crops they have grown.

The themes of recovery, repatriation, and conservation of traditional crops are of mutual interest to indigenous leaders and international donors. Biodiversity has become a global priority with considerable funding, a fact not lost on indigenous leaders. During our research in Cotacachi (2000–2006), at any one time as many as six biodiversity projects were operating simultaneously. These projects, each with its own donor, typically functioned independently of each other and often several worked within a single village with the same local collaborators. Indigenous leaders have even been known to successfully submit the same proposal to several different agencies.

The types of crop biodiversity conservation and utilization projects in Cotacachi are typical of those found throughout the developing world. For example, one project is called Ally Tarpuy, which is Quichua for "good planting" (Spanish: *buena siembra*). It is sponsored by the European Union and run by an Italian NGO with the purpose of supporting school biodiversity gardens and agroecology demonstration farms using organic methods. Another project is La Red de Guardianes de Semillas (Network of Seed Keepers), and one of its main objectives is to sponsor an annual "festival of Andean grains and seeds" where participating communities bring seeds to exchange. A much more ambitious project, titled Complementary Conservation and Sustainable Use of Under-utilized Crops in Ecuador, is implemented by USDA and Instituto Nacional Autonomo de Investigaciones Agropecuarias (INIAP; Autonomous National Institute of Agricultural Research) and its national genebank draws on PL480 debt funds (INIAP and USDA 2002). With more than a half million U.S. dollars from PL480, this thirty-month project (October 2002–March 2005) aimed to collect specific germplasm, promote its use, improve well-being of communities through food security, strengthen the connection between genebanks and *in situ* conservation, and enhance germplasm that is underrepresented in the genebanks.

While all biodiversity recovery projects in Cotacachi funnel money into the local economy and raise awareness of traditional crops, it remains unclear whether local priorities in biodiversity are being addressed. Outside organizations, whether international agencies, governmental organizations, or NGOs, bring their own agendas, which often appeal more to donors and scientists than to local people. The PL480 project mentioned above focused much of its work on three crops: *tomate de árbol* (tree tomato; Spanish: *zapallo*), Andean squash, and *aji* (a hot pepper variety). These crops were prioritized by national genebank scientists for their commercial potential and "the national good." One internationally known NGO that specializes in animals has been "repatriating" llamas and alpacas to Cotacachi when in fact Andean cameloids never existed in the Cotacachi area. Another NGO project aimed to combine ecotourism and biodiversity conservation through creation of ethnic gardens near village eco-hostels. The gardens were envisioned as foregrounding traditional landraces, thus encouraging *in situ* conservation, as well as raising incomes for indigenous families through cultural tourism. However, either through some miscommunication or conflicting project objectives, NGO and expatriate advisers assisted villagers in planting mainly European materials such as lettuce, broccoli, Swiss chard, and onions. The justification for the predominance of Old World crops over Andean ones was that the gardens should produce vegetables palatable to tourists' tastes when they take meals at the hostel.

Despite lip service to ethnic identity and place, the primary objectives of many outside programs are to conserve biodiversity through income enhancement and scientific management. However, as with American Southerners and Vietnamese immigrants, the primary motivations of gardeners and farmers in Cotacachi for species recovery and use are cultural and tied to memory, place, and ethnicity (see chapter 5, this volume). Some poststructural anthropologists have also argued that "biodiversity" in situations like Cotacachi's should be understood as "a discursive invention of recent origin," a nascent product of a global network of actors, comprising international agencies, NGOs, scientists, bioprospectors, local communities, and social movements (Escobar 1998:53). The latter two, it is argued, couch biodiversity in terms of "cultural difference, territorial defense and social and political autonomy" (Escobar 1998:54). While this "discourse" captures much of what is happening in Cotacachi, such an interpretation foregrounds the sociopolitical world of outsiders and indigenous intellectuals, not village-level gardeners and farmers. Cotacacheños, through centuries of political and environmental disruptions, have maintained a rich complex of native crops (see figure 12.3). While they see a strong link between territoriality and their agriculture, they are realistic today that agriculture is insufficient to support their complete economic needs. Agroeconomic and anthropological research in Cotacachi shows that farming, especially maize and bean cultivation, is motivated more by ceremonial and basic subsistence uses than market income (Rhoades 2006b). Both the neoliberal "development" interpretation of agencies and the views of poststructural anthropologists seem to deny this agency of everyday people who are in the main unaware of liberation discourse, science, or global markets for their crops.

While development practitioners and indigenous leaders see hope that their Andean crops can be linked to emerging niche markets, access to these markets is poorly understood and still remote for local people.

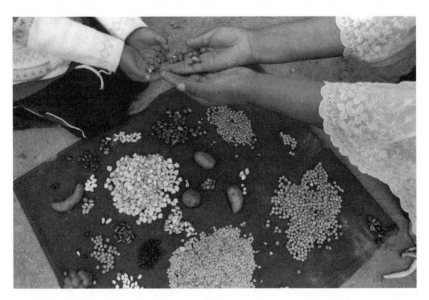

FIGURE 12.3. Seeds and tubers pass from mother to child in Cotacachi, Ecuador. Photo by Robert E. Rhoades.

The indigenous leadership is becoming aware that mestizos and foreigners are keenly interested in Andean grains, such as quinoa, which is becoming a profitable export crop. Increasingly, international NGOs and national programs, including INIAP, are setting up projects to capture foreign niche markets, including those generated by the global organic, health food movement. In the meantime, the role of culture, place, and memory as primary motivators for local preservation of biodiversity in marginal Andean communities should not be neglected. The traditional food habits and the Andean crops as cultural symbols are deeply ingrained in the Cotacacheños psyche and are central to their historical resistance against assimilation into either mestizo or global cultures. The knowledge of how to produce and prepare these native crops still survives among the elderly and poorer, more remote households. They guard the increasingly scarce landraces in gardens and rocky fields without support or recognition from the outside.

Conclusion

In the growing literature on globalization and its impacts on biodiversity, the agency and cultural goals of local people have received the least attention. Formal organizations—genebanks, international organizations, NGOs, government organizations—justify their activities on this theme largely in terms of the emerging global market economy and managerial efficiency. While recognizing the need for the well-being of local people, these external formulations still envision the recuperation of lost or vanishing crops as a process that must be scientifically informed, expertly managed, and economically viable in light of new global market realities. This chapter, however, has looked at three cultures in which the majority of local, informal seedsavers curate their culturally relevant plants with none of these goals in mind. Outside a select few indigenous leaders in Cotacachi, Ecuador, very few grassroots gardeners or farmers among American Southerners, Vietnamese immigrants, or Cotacacheños are aware of how outsiders interpret their passion for recuperating and saving historical plants.

In all three societies, globalization of the past five decades has sadly wiped out some of their rich and distinct crop diversity and the associated agricultural knowledge and culinary practices. Although the circumstances of the three cases vary, all have experienced new lifestyles brought on by a global expansion of markets and communications, population displacements, and exposure to new foods and diets (Phillips 2006). The lifestyles required by a modern, globalized world hardly leave time to nurture old-timey crops, let alone process them for the kitchen and then cook them into traditional dishes. However, the three cases also demonstrate that we are myopically looking at the glass as "half empty" if we see only destructive forces operating on small-scale farmers and gardeners around the world. A "half-full" vision can also be recognized in which stubborn, independent, and driven seedsavers refuse to give up their heirloom plants or their plants from the homeland.

The three marginalized groups in this chapter, as well as communities and cultures in other geographical locations included in this volume, creatively seek out landraces and nurture them within social and cultural contexts driven by memory, identity, and sense of place. Much of the activity around seeds is a strong indication of agency and reterritorialization played out on the colorful stage of everyday community, kinship, ceremony, ritual, and gathering around food. The intensity and magnitude of thousands of officially unseen seedsavers in seeking, reviving, and exchanging traditional planting materials cannot be written off as economically naive nostalgia or epiphenomenal "discourse." Their biodiversity conservation efforts "out there" are equally significant, if not superior, to the global efforts of scientists, seed banks, and conservation projects.

References

Airriess, Christopher A., and David L. Clawson. 1991. Versailles: A Vietnamese enclave in New Orleans, Louisiana. *Journal of Cultural Geography* 12:1–13.

Appadurai, Arjun. 2001. *Globalization*. Durham, NC: Duke University Press.

Bebbington, Anthony. 1996. Movements, modernizations, and markets. Indigenous organizations and agrarian strategies in Ecuador. In *Liberation Ecologies: Environment, Development, Social Movements*, edited by R. Peet and M. Watts, 86–109. Routledge: London.

———. 2001. Globalized Andes? Livelihoods, landscapes and development. *Ecumene* 8(4):414–436.

Bender, Steve, and Felder Rushing. 1993. *Pass-Along Plants*. Chapel Hill, NC: University of North Carolina Press.

Calhoun, Creighton Lee. 1995. *Old Southern Apples*. Blacksburg, VA: McDonald and Woodward.

Camacho, Juana. 2006. Good to eat, good to think: Food, culture and biodiversity in Cotacachi. In *Development with Identity: Community, Culture and Sustainability in the Andes*, edited by Robert E. Rhoades, 156–172. Cambridge, MA: CABI.

Davis, Edward, and John Morgan. 2005. Collards in North Carolina. *Southeastern Geographer* 45(1):67–82.

Do, HienDuc. 1999. *The Vietnamese Americans*. Westport, CT: Greenwood Press.

Do, Lynn. 2003. Vietnamese homegardens in Southwest Florida. In *Introduced Germplasm from Vietnam: Documentation, Acquisition, and Preservation,* edited by Robert E. Rhoades and Virginia D. Nazarea, 32–67. Athens, GA: Department of Anthropology.

Escobar, Arturo. 1998. Whose knowledge, whose nature? Biodiversity, conservation, and the political ecology of social movements. *Journal of Political Ecology* 5:53–82.

Gepts, Paul. 2006. Plant genetic resources conservation and utilization: The accomplishments and future of a societal insurance policy. *Crop Science* 46:2278–2292.

Hurt, R. Douglas, ed. 1998. *The Rural South since World War II*. Baton Rouge: Louisana State University Press.

INIAP and U.S. Department of Agriculture. 2002. *Conservación Complementaria y Uso Sostenible de Cultivos Subutilizados en Ecuador*. Quito, Ecuador: Instituto Nacional Autonomo de Investigaciones Agropecuarias.

Kalcik, Susan. 1984. Ethnic foodways in America: Symbol and the performance of identity. In *Ethnic and Regional Foodways in the United States: The Performance of Group Identity*, edited by Linda K. Brown and Kay Mussel, 37–65. Knoxville: University of Tennessee Press.

Moates, A. Shiloh, and Brian C. Campbell. 2006. Incursion, fragmentation and tradition: Historical ecology of Andean Cotacachi. In *Development with Identity: Community, Culture and Sustainability in the Andes*, edited by Robert E. Rhoades, 24–45. Cambridge, MA: CABI.

National Research Council. 1989. *Lost Crops of the Incas: Little-Known Plants of the Andes with Promise for Worldwide Cultivation*. Washington, DC: National Academy Press.

———. 1998. *Cultural Memory and Biodiversity*. Tucson: University of Arizona Press.

Nazarea, V. D. 2005. *Heirloom Seeds and Their Keepers: Marginality and Memory in the Conservation of Biological Diversity*. Tucson, AZ: University of Arizona Press.

———. 2006. Local knowledge and memory in biodiversity. *Annual Review of Anthropology*. 35:317–335.

Ogle, Britta M., Pham H. Hung, and Ho To. Thuyet. 2001. Significance of Wild Vegetables in Micronutrient Intake of Women in Vietnam. *Asia Pacific Journal of Clinical Nutrition* 10(1):21–30.

Owens, Richard. 2003. Vietnamese home gardens of Lincoln, Nebraska: A measurement of cultural continuity. Master's thesis, Department of Anthropology, University of Nebraska, Lincoln.

Peacock, James L., Harry L. Watson, and Carrie R. Mathews, eds. 2005. *The American South in a Global World*. Chapel Hill, NC: University of North Carolina.

Péralte, Paul C. 2003. Seminole pea serves hope for prosperity. *Atlanta Journal Constitution*, January 1.

Phillips, Lynne. 2006. Food and globalization. *Annual Review of Anthropology* 35:37–57.

Pillsbury, Richard. 2006. *The New Encyclopedia of Southern Culture*, Vol. 2: *Geography*. Chapel Hill, NC: University of North Carolina Press.

Piniero, Maricel. 2006. Women and homegardens of Cotacachi. In *Development with Identity: Community, Culture and Sustainability in the Andes*, edited by Robert E. Rhoades, 140–155. Cambridge, MA: CABI.

Rhoades, Robert E., ed. 2006a. *Development with Identity: Community, Culture and Sustainability in the Andes*. Cambridge, MA: CABI.

———. 2006b. Linking sustainability science, community and culture: A research partnership in Cotacachi, Ecuador. In *Development with Identity: Community, Culture and Sustainability in the Andes*, edited by Robert E. Rhoades, 1–15. Cambridge, MA: CABI.

Rhoades, Robert E., and Virginia D. Nazarea. 1998. Local management of biodiversity in traditional agroecosystems: A neglected resource. In *Importance of Biodiversity in Agroecosystems*, edited by Wanda Collins and Calvin Qualset, 215–236. Boca Raton, FL: Lewis/CRC Press.

———. 2003. Introduced germplasm from Vietnam: Documentation, acquisition and preservation. Report submitted to U.S. Department of Agriculture/Agricultural Research System.

Shortridge, B., and J. Shortridge, eds. 1998. *The Taste of American Place*. Lanham, MD: Rowan and Littlefield.

Skarbø, Kristine. 2006. Living, dwindling, finding: Status and changes in agrobiodiversity of Cotacachi. In *Development with Identity: Community, Culture and Sustainability in the Andes*, edited by Robert E. Rhoades, 123–139. Cambridge, MA: CABI.

Wibbelsman, Michelle. 2005. Encuentros: Dances of the Inti Raymi in Cotacachi, Ecuador. *Latin American Music Review* 26(2):195–226.

CONTRIBUTORS

JENNA E. ANDREWS-SWANN is an environmental anthropologist whose research focuses on place making in the politicized context of immigration. She has explored the intricate ways that nostalgia and sensory memories can contribute to the production of transnational landscapes: those lived-in places that link and (re)create particular elements of home and host country. Much of her work deals with these issues in urban and periurban spaces throughout the United States, locales often overlooked by anthropologists. Andrews-Swann is assistant professor of anthropology at Georgia Gwinnett College near Atlanta, Georgia.

PETER BROWN is a public school teacher in San Diego, California who has actively supported the Zapatista Education System of Chiapas, Mexico since 1996. Brown also serves as an elected delegate to the Representative Assembly of the National Education Association and is a coordinator of Schools for Chiapas/Escuelas para Chiapas.

TOM BROWN of Clemmons, North Carolina, became interested in finding and saving old-timey apples in 1999. His efforts have resulted in the relocation, propagation, and repatriation of over nine hundred of these varieties. He also carefully collects stories about each apple's heritage and use to share with fellow apple enthusiasts and was a recipient of the 2008 Colporteurs-in-Residence Award from the Fostering Our Local Knowledge Group and the Ethnoecology/Biodiversity Laboratory. Brown's work continues at www.applesearch.org.

JUANA CAMACHO is a researcher at the Instituto Colombiano de Antropologia e Historia in Bogotá, Colombia. She served as editor for the *Revista Colombiana de Antropologia* from 2007 to 2009. Camacho has worked extensively on environmental and social issues with black communities in the Colombian Pacific coast and is currently studying the food traditions and agricultural practices of peasant communities in the Colombian Andes.

SUSANNAH CHAPMAN is currently writing up her research on The Gambia focusing on Mandinka farmers' understandings and perceptions of the concepts embedded in intellectual property rights law for plants as it is currently drafted under international law. Her research combines ethno-ecology, political economy, political ecology, and agricultural anthropology to explore issues raised by the expansion of intellectual property rights law and emerging agricultural development policy. Chapman is the recipient of a National Science Foundation doctoral dissertation research improvement grant, and she has coauthored several articles on apple diversity in the United States.

CARY FOWLER is the executive director of the Global Crop Diversity Trust. He received the Vavilov Medal for his contribution to the cause of conserving plant genetic resources and the Heinz Award for his vision and efforts in the preservation of the world's food supply. He headed the International Conference and Programme on Plant Genetic Resources at the U.N. Food and Agriculture Organization (FAO), which produced the first U.N. global assessment of the state of the world's plant genetic resources, and supervised negotiations of FAO's Global Plan of Action for Plant Genetic Resources. Currently, he serves as chair of the International Advisory Council of the Svalbard Global Seed Vault.

MAGDALENA FUERES, RODRIGO FLORES, and ROSITA RAMOS are leaders of the Union of Peasant and Indigenous Organizations of Cotacachi (UN-ORCAC), a nonprofit organization that joins forty-one communities, including peasants, indigenous people, and farmers in the Andean zone of Cotacachi canton, Imbabura province, Ecuador. UNORCAC was created in 1977, after a long organizational process that was headed by a group of young indigenous Cotacachi intellectuals. The desire to eliminate the conditions of discrimination and poverty, which the majority of rural and indigenous peoples faced in the area, was the main motivation for uniting the communities. UNORCAC is affiliated with the Federation of Indigenous Peasants of Imbabura (FICAPI) and the National Federation of Indigenous, Peasant, and Black Organizations (FENOCIN).

TIRSO GONZALES is Peruvian of Aymara descent. As scholar, international consultant, and activist, he works closely with concerns and issues in relation to indigenous peoples throughout the Americas. He is currently on the faculty of Indigenous Studies at the University of British Colombia–Okanagan in Canada and is a member of the British Columbia Food Systems Network Working Group on Indigenous Food Sovereignty. He was a member of the Peruvian National Commission of Indigenous Andean,

Amazonian, and Afro Peruvian People. Gonzales's recent research explores the use of participatory methodologies and techniques to address problems central to indigenous strategic visions and local management of natural resources with a firm commitment to cultural affirmation and decolonization.

RICHARD MOORE is a recent president of the Culture and Agriculture Section of the American Anthropological Association. He is professor and director of the Environmental Science Graduate Program and serves as assistant director of the School of Environment and Natural Resources at Ohio State University. He is also an adjunct professor in anthropology. Moore has served as lead principal investigator on grants from the National Science Foundation, U.S. Agency for International Development, and U.S. Environment Protection Agency, principally with themes on the interface of farming and water quality. Moore has published widely on these themes, and he is also a recipient of the Carnegie Foundation Award for Outreach and Engagement.

VIRGINIA D. NAZAREA's research has focused on the situatedness of local knowledge and its distribution according to ethnicity, age, class, and gender (*Local Knowledge and Agricultural Decision Making: Class, Gender, and Resistance*, 1995; *Ethnoecology: Situated Knowledge/Located Lives*, 1999). Nazarea is the recipient of the Creative Research Award for the Social Sciences, the Scholarship of Engagement Award in 2009, and the National Endowment for the Humanities Fellowship. In her work she examines how memory and marginal spaces allow women, small-scale farmers, and indigenous people to nurture a diversity of plants that otherwise may be lost to agricultural commercialization and monoculture (*Cultural Memory and Biodiversity*, 1998; *Heirloom Seeds and Their Keepers: Memory and Marginality in the Conservation of Biological Diversity*, 2005).

ROBERT E. RHOADES was Distinguished Research Professor of Anthropology and director of the Sustainable Human Ecosystems Laboratory at the University of Georgia. With more than four decades of experience in international development and conservation, he has authored more than 160 publications with a focus on agrobiodiversity-related problems and solutions. He has won numerous awards for his research and publications, including a National Writer's Award, Praxis Award in Anthropology, and Fulbright Fellowship. He served on the U.S. Genetic Resources Advisory Board, cofounded the Southern Seed Legacy, and was named the Calvin Sperling Memorial Distinguished Lecturer by the Crop Society of America.

KRISTINE SKARBØ is a researcher at Vestlandsforsking in Norway. She has done ethnographic research in the areas of food, agriculture, and biodiversity from Uganda, Ecuador, Bolivia, and Peru. She has completed her dissertation research on reindigenization, food, and conservation of agrobiodiversity in the Ecuadorian Andes with funding from the National Science Foundation. She has published on a variety of topics, including anthropological contributions to biodiversity and indigenous conceptions of climate change.

JAMES R. VETETO is an environmental anthropologist specializing in ethnoecology, agrobiodiversity studies, sustainable agricultural systems, sustainable development, food and culture, and ecotopian possibilities. He is assistant professor and director of the Laboratory of Environmental Anthropology at the University of North Texas and current director of the Southern Seed Legacy. Veteto has worked extensively with local and indigenous communities in southern Appalachia, the Ozarks, and the Sierra Madre Mountains of Northwest Mexico. His work has focused specifically on comparative agrobiodiversity inventories, farmer decision making, and conservation strategies in mountain ecosystems.

KEVIN WELCH is a member of the Eastern Band of Cherokee Indians and lives in the Big Cove Community of Cherokee, North Carolina, USA. He is the founder and coordinator of the Center for Cherokee Plants and FRTEP (Federally Recognized Tribes Extension Program) assistant for the Eastern Band of Cherokee Indians Cooperative Extension. In 2005, Welch initiated the Cherokee Traditional Seeds Project that led him to create the Center for Cherokee Plants, a tribal seed bank and native plant nursery. He has introduced two important resolutions for tribal legislation to protect Cherokee intellectual property rights and native plants from exploitation and destruction.

INDEX

Page numbers in italic indicate illustrations.